# Roofing

# Management

*An Overview of the Roofing Industry*

Developed with funding provided by
The Roofing Alliance

**ROOFING
ALLIANCE**
THE FOUNDATION OF NRCA

*Nieri Department of*
**CONSTRUCTION,
DEVELOPMENT AND
PLANNING**

© 2025 Nieri Department of Construction and Real Estate Development and the Roofing Alliance

ISBN 978-1-63804-184-9
eISBN 978-1-63804-185-6

Distributed by Clemson University Press

# Roofing Management

*An Overview of the Roofing Industry*

Dr. Dhaval Gajjar

Dr. Jason Lucas

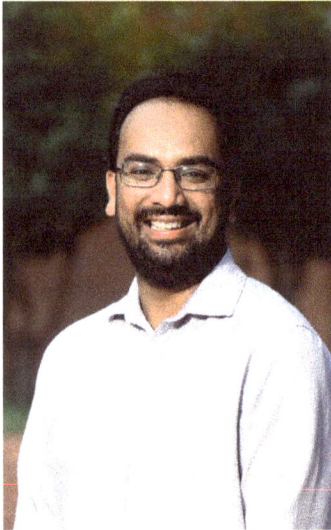

## Dhaval Gajjar, Ph.D., FMP, SFP

Dr. Dhaval Gajjar is an Associate Professor at Clemson University's Department of Construction and Real Estate Development in the College of Architecture, Arts and Construction. Dr. Gajjar has significant construction industry experience working as a Project Manager for both the general contractor and an owner organization responsible for large remodel, renovation, and IT projects. Dr. Gajjar also has 10+ years of significant research experience related to workforce development, education and training, project delivery and performance measurement. He has authored over twenty (20+) refereed journal publications, one book chapter, and over fifteen (15+) conference presentations disseminating the research results. He has also conducted over fifty (50) industry presentations educating industry professionals on using the research tools. He is also a recipient of Bennett Award and is a certified Facility Management Professional (FMP) and Sustainable Facility Professional (SFP).

## Jason Lucas, Ph.D.

Dr. Jason Lucas is an Associate Professor in the Department of Construction, Development and Planning at Clemson University. He holds a Bachelor of Architecture degree from the New Jersey Institute of Technology and a master's in Building Construction Science and Management and a PhD in Environmental Design and Planning from Virginia Tech. Dr. Lucas has conducted research in workforce development, online education, and the use of technology in education. He has over 10 years of research and teaching experience at Clemson University and has published over twenty (20+) journal articles and over thirty (30+) conference presentations.

# Acknowledgements

### Assistant Editor

## Gopika Viswanathan

Gopika Viswanathan is a graduate student in the Masters of Construction Science and Management program. She also works as a Research and Teaching Assistant at Clemson University.

### Layout Designers

## Danett Vargas Sanchez

Danett Vargas Sanchez is an undergraduate student double majoring in Art and Communication at Clemson University. She is part of the Clemson University Honors College and has four years of graphic design and fine art experience.

## Amelia Lyles

Amelia Lyles is an undergraduate student majoring in Graphic Communications at Clemson University. She also works as a Teacher Assistant for Clemson and has two years of graphic design experience.

### Technical Reviewers

**William Good,** Roofing Alliance
**Mark Graham,** NRCA

**Alison L. LaValley,** CAE
**Jessica Priske,** Director
**Nicole Christodoulou,** Manager
**Maggie Kosinski,** Manager

# Acknowledgements

Contributors:

**Jennifer Ford-Smith,** Johns Manville

**Billy Jones,** CDC, Inc.

**Mark S. Graham,** National Roofing Contractors Association

**Helene Hardy Pierce,** Industry Consultant

**Steve Kruger,** L.E. Schwartz & Son, Inc.

**Dr. Anne Cope, P.E.,** Insurance Institute for Business & Home Safety

**Brad Sutter,** Sutter Roofing

**Curt Boyd,** Academy Roofing, Inc.

**Rick Damato,** Damato Enterprises

**Glenn Watso,** Advanced Roofing

This series of manuals on the roofing industry is made possible through the support of the Roofing Alliance and numerous industry professionals. The manuals were developed from lectures given by industry experts and are meant as an introduction to the topics. It is not intended to include the entirety of the industry and the information about all types of roofing products or systems. For a more complete discussion of roofing materials and installation methods, the reader is encourage to see the NRCA Roofing Manual, available from the National Roofing Contractors Association (www. nrca.net).

The information contained has been reviewed for technical accuracy and clarity at the date of its publication. Codes and practices change over time, so the editors intend to periodically review, revise, and publish future editions of these manuals to reflect those changes.

The roofing manual series can be used independently to provide an overview of different parts of the roofing industry but also serve as a complimentary summary of the knowledge presented in the Clemson Online Professional Development series of courses that have been created with the support of the Roofing Alliance:

- Roofing Fundamentals
- Roofing Management
- Roofing Business and Leadership

The content is broken up into three topic areas to allow tailored focus of employees and business leaders to focus on the area most relevant for them. Roofing Fundamentals provides a general overview of the industry, the products and services available, and focuses on developing an understanding of systems and terminology. Roofing Management covers topics related to a project, including codes, scheduling, field crew management, quality control, risk management, and site logistics. Finally, Roofing Business and Leadership delves into leadership strategies, sales, marketing, and various aspects of owning a business within the roofing industry.

This project would not have been possible without the numerous industry supporters who have donated time and content in support of creating these educational resources.

# Table of Contents

Chapter Name                                                              Page Number

## Chapter 1: Commercial Roofing                                              1

1.1 Commercial Roofing Market                                                  2

1.2 The Roof as a System                                                       2

1.3 Wind and Roofing                                                           3

1.4 Fire Conditions                                                            4

1.5 Hail Considerations                                                        5

1.6 Energy Efficiency                                                          6

1.7 Product Selection                                                          8

1.8 Insulation and Coverboards                                                12

1.10 Accessories and Fasteners                                                14

## Chapter 2: Building Codes in Roofing                                      16

2.1 Introduction                                                             17

2.2 The Contractor's Responsibility                                          19

2.3 Building Codes and Standards                                             19

2.4 Wind and Load Requirements                                              22

2.5 Fire Resistance Requirements                                            23

2.6 Roof Accessory Requirements                                            24

2.7 Prescriptive Requirements for Roof Assemblies                          25

2.8 Reroofing Requirements                                                  26

2.9 Residential Building Code and Existing Building Code                    27

2.10 Energy Code Requirements                                              27

2.11 Plumbing Code                                                          28

2.12 Fire Safety                                                            28

2.13 Reference Standards                                                    29

## Chapter 3: The Roofing Industry and the Environment     31

3.1 The Roofing Industry and the Environment     32

3.2 Rating Systems     33

3.3 Resiliency and Roofing     39

3.4 Resilient Designs     40

3.5 Impact Resistance     41

3.6 Daylighting     42

3.7 Roof Insulation and Resiliency     42

3.8 Considerations for Solar Energy Production     44

## Chapter 4: Roof Scheduling     47

4.1 Introduction     48

4.2 Scheduling Considerations     51

4.3. Schedule Coordination     52

4.4 Creating a Roof Schedule     52

4.5 Scheduling Example     53

## Chapter 5: Insurance and Risk Management     58

5.1 Insurance     59

5.2 Risk Management     61

## Chapter 6: Site Logistics     64

6.1 Introduction     65

6.2 In-bound and out-bound operations     67

6.3 Job Sequencing     69

6.4 Safety Logistics     71

# Table of Contents

Chapter Name                                    Page Number

## Chapter 7: Field Crew Management              72

7.1 Crew Structure                               73

7.2 Crew Planning and Monitoring                 73


## Chapter 8: Roofing Equipment                  77

8.1 Introduction                                 78

8.2 Safety Equipment                             79

8.3 Reroofing Equipment                          82

8.4 Material Handling Equipment                  85

8.5 Built-up Roofing Equipment                   87

8.6 Modified Bitumen Equipment                   91

8.7 Spray Foam and Coating Equipment             92

8.8 Single-Ply Roofing Equipment                 93

8.9 Metal Roofing Equipment                      95

8.10 Residential Roofing Equipment               97

8.11 Slater's Tools                              97

8.12 Miscellaneous Tools and Gadgets             98

8.13 The Future of Roofing Equipment             99

# Table of Contents

Chapter Name                                                    Page Number

**Chapter 9: Quality Control and Quality Assurance**            **100**

9.1 Introduction                                                101

9.2 The QA/QC Team                                              103

9.3 Standard of Work                                            104

9.4 Typical Repairs                                             107

9.5 Continous Personnel Improvement                             110

9.6 Software                                                    110

9.7 Testing                                                     111

# CHAPTER

# 01

Commercial Roofing

With contributions by
Jennifer Paul Smith
Usher Manville Corp

Companies operating in the commercial roofing industry provide roofing services and materials for such commercial buildings as schools, retail stores, warehouses and hospitals. Commonly used commercial roofing materials include TPO (thermoplastic polyolefin), PVC (polyvinyl chloride), EPDM (ethylene propylene diene monomer), asphalt and modified bitumen.

Figure 1.1.1 shows material usage in the commercial roofing industry from 2001 to 2019. TPO has grown significantly, today accounting for more than 45% of the market. This is primarily due to its low cost and ease of installation.

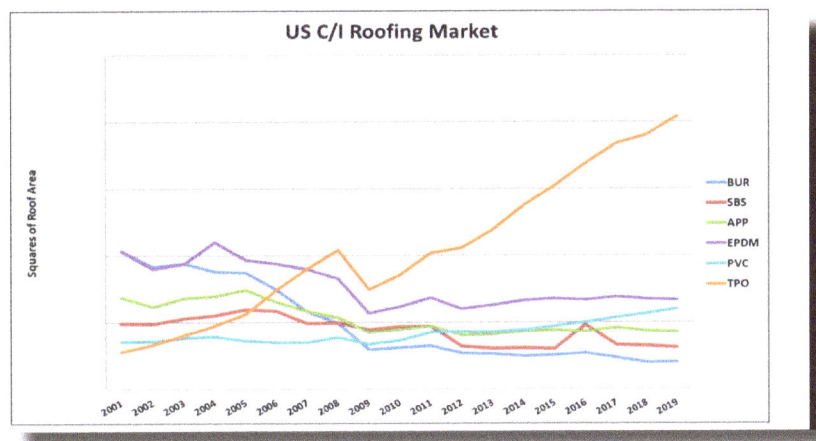

Figure 1.1.1 - Roofing Market Data

## 1.2 The Roof as a System

A high-performing roof system requires four elements:

- A detailed design
- Products that perform
- Quality installation
- Maintenance

Most roofing failures can be attributed to poor design and/or poor installation.

Roof specifications are essential to ensure the roof is installed correctly and performs for at least its intended life. Wind, water and fire resistance requirements, the interior environment of the building, the purpose of the building, the exterior environment and durability all contribute to the roof's specifications.

Many people influence the type of roof system that is selected. These may include architects, designers, distributors, manufacturers, consultants and roofing contractors. An illustration of the various entities involved is shown in Figure 1.2.1.

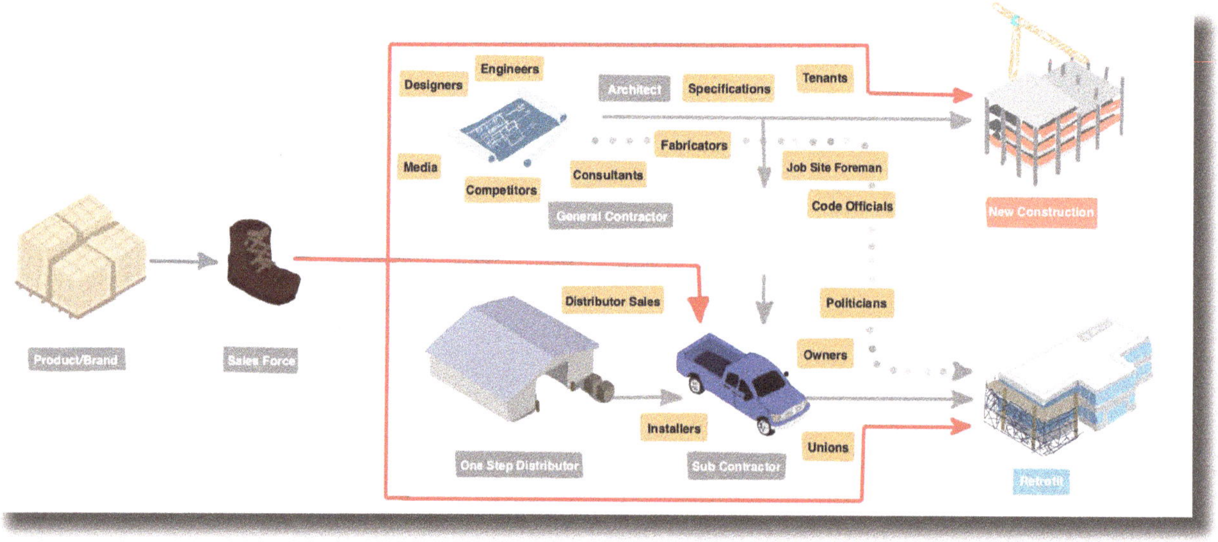

Figure 1.2.1 – Product Selection

## 1.3 Wind and Roofing

Building code requirements in coastal areas that experience frequent storms are much more rigid than in other parts of the U.S. In addition, there are other bodies involved in developing wind standards for roofing, most notably Factory Mutual Approvals (FM). FM is an insurance provider that offers insurance for commercial buildings only if the roof meets its testing requirements. While only about 10% of buildings in the U.S. are insured by FM Global, its requirements have become a standard in the industry. Specifications often include terms such as "FM 1-90," indicating the wind uplift pressure that the roof system must be designed to withstand.

Other bodies setting wind standards for roof systems include the Miami-Dade (Florida) building code, Underwriters Laboratories (UL), Directory of Roof Assemblies (DORA) and Texas Department of Insurance.

Proper edge metal design and installation has its own set of requirements. Leaks at perimeters and corners are common problems that can lead to roof system failure. The edge metal design must conform to what is known as ANSI/SPRI ES-1, developed by SPRI, an association representing single-ply roofing manufacturers. A roof collapse can cause serious damage and injury.

## 1.4 Fire Conditions

Strict fire codes were first developed in the aftermath of the Great Chicago Fire of 1871. It took another 10 to 15 years for fire-resistant building materials to evolve, and ultimately, such materials as polyisocyanurate roof insulation had success due, at least in part, to their fire-resistant properties.

Figure 1.4.1 shows a "burning brand" test conducted by UL, where the roof system is sloped and set on fire. The distance the roof system burns during a set time will determine whether it is rated Class A, B or C for fire resistance, with Class A being the most desirable. With most materials, the rating decreases as the slope increases. Different materials have different fire-resistant properties. For example, PVC membranes are more fire-resistant than TPO membranes.

Figure 1.4.1– UL Burning Brand Test

## 1.5 Hail Considerations

Hail is a major cause of roof damage. There has been a 65% increase in hail claim severity since 2007, and hail now accounts for about 40% of insurance losses for homes and buildings. Hail ratings in the U.S. are classified as:

- Moderate hail

- Severe hail

- Very severe hail

Moderate hail approval is for hail up to 1"; severe hail is for hail up to 1 ¾"; very severe hail approval is for hail greater than 2".

There are different standards associated with hail ratings, including those developed by the American Society for Testing and Materials International (ASTM), UL and FM. The FM standard is the one most frequently referenced.

Figure 1.5.1 shows a roof membrane damaged by hail. This is a potential leak source. In many cases, the hail does not necessarily puncture or crack the membrane but can compress the roof insulation under it, resulting in a reduction of the insulation's R-value.

Figure 1.5.4 Hail Damage

One strategy to help protect against hail damage is to use a rigid substrate, which can be achieved by installing a dense cover board over the insulation. Another strategy is to use materials that increase energy absorption; these include EPDM and fleece-backed membranes.

A building loses most of its energy through the windows and the roof. Therefore, the design of the roof must take into account its energy efficiency.

Some building codes specify the use of a white (or other reflective) roof since it is more likely to reflect sunlight and reduce heat gain. The International Energy Conservation Code (IECC) and standards developed by the American Society of Heating, Refrigerating and Air-Conditioning Engineers (ASHRAE) are commonly used in the industry for specifying energy-saving roofs. Building codes typically require a roof system to have a minimum R-value that will depend on which of the eight climate zones the building is located in.

There is continuing debate in the industry about whether energy conservation is best achieved by using more insulation or by using materials that are highly reflective and/ or emissive. In areas of the country with more heating days and fewer cooling days, for example, reflectivity becomes less important, and heat-absorbing materials may be more energy-efficient.

As a general rule, light-colored roofs are more popular in warmer climates, and dark-colored roofs are more popular in colder climates. An example of a reflective TPO roof is shown in Figure 1.6.1.

Figure 1.6.1–Cool Roof

Another roof system that can help save energy is a vegetative or "green" roof system, as shown in Figure 1.6.2. Vegetative roof systems can range from a simple design, such as using modular or tray systems, to more complicated designs where the vegetation is made part of the system. Modular or tray systems are the most popular due to the comparative ease of finding and repairing leaks or defects in the system.

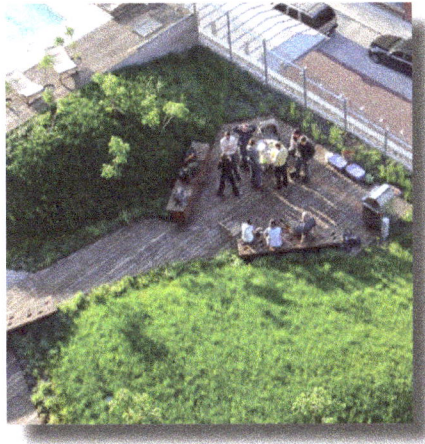

Figure 1.6.3 – Vegetative Roof

One other important design consideration is accounting for rooftop equipment, vents and mechanical units. Too often, these are affixed directly to the roof membrane, puncturing it. Rooftop equipment should be mounted on curbs. All penetrations should be flashed to prevent leaks and keep the roof system watertight. Penetrations should be round or square and easy to flash.

A low-slope roof typically consists of a roof deck, a vapor retarder, thermal insulation and a membrane. It may also include a cover board and surfacing material (See Figure 1.7.1). When a cover board is used, it is applied on top of the insulation with the membrane on top of it, attached to the assembly.

The following are the most common types of low-slope roof systems:

Built-up roofs (BUR)is a multi-layer system made by embedding multiple layers (2-4) of felt in asphalt, usually with a gravel surface. Built-up roofs are considered to be very durable and able to withstand harsh weather conditions. The process of installing a BUR is labor-intensive, and asphalt emits an unpleasant odor when it is heated. BUR has lost market share in recent years.

SBS modified bitumen roofsare composed of pre-manufactured rolls of felt and bitumen modified with a polymer called styrene butadiene styrene or SBS. It is a two-ply system that works well in cold climates and has an excellent history of performance in high-wind areas. SBS modified bitumen roofs can be applied using torches, hot asphalt and adhesives; some products are also self-adhering. SBS modified bitumen systems do not work well when they are exposed to oils or hydrocarbon-based materials.

Atactic Polypropylene (APP) is another polymer modified used to make modified bitumen roof systems.

An example is shown in Figure 1.7.2. APP has more plastic-like properties than SBS, which tends to be more rubbery. APP is best used in high temperatures and is usually applied using a torch. Torch applications require great care, and some insurance companies require the worker using the torch to be a certified applicator. NRCA and other roofing organizations offer certification training programs under the brand name "CERTA."

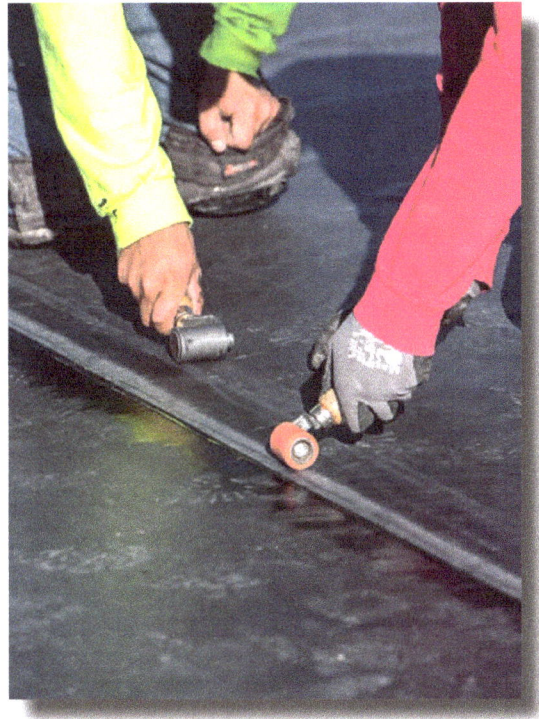

 Ethylene Propylene Diene Monomer (EPDM) is a type of synthetic rubber, as shown in Figure 1.7.3 that has been in the roofing market for about 60 years. It is typically black but is also made in white to improve reflectivity. EPDM has good resistance to cold weather and is UV stable. Seams can sealed with adhesive or can be heat welded. EPDM does not work well with oils and needs attention paid to priming, taping and seam application when installed.

As shown in Figure 1.7.4, thermoplastic olefin (TPO) is a relatively new, highly reflective roofing material. It is suitable for warm climates and is usually a cost-effective option. It is easy to install, durable and energy efficient. TPO membranes do not handle oils well, and cold seaming can be difficult.

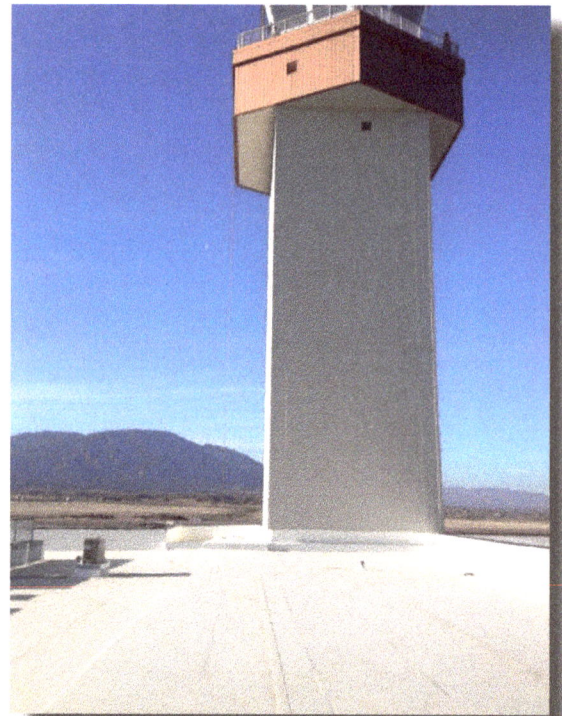

Figure 1.7.4—TPO Roofs                    Figure 1.7.5—PVC Roofing

As shown in Figure 1.7.5, polyvinyl chloride (PVC) roof membranes come in a single layer and contain both polyvinyl chloride and plasticizers. PVC has excellent resistance to oil and is commonly seen on the roofs of airports, manufacturing plants and restaurants. PVC is generally softer and easier to weld than TPO. It is typically white in color and is highly reflective and energy efficient.

As shown in Figure 1.7.6, liquid-applied roofs are becoming more popular and are commonly used for plaza decks, vegetative roof systems, areas that are difficult to access, irregular roofs and small areas with a substantial number of penetrations. Liquid-applied roofs provide a seamless application and are easy to install. They can also be used to coat existing roof systems, which can be much less expensive than buying a new roof. These roofs can be applied by spraying, brushing or using a roller.

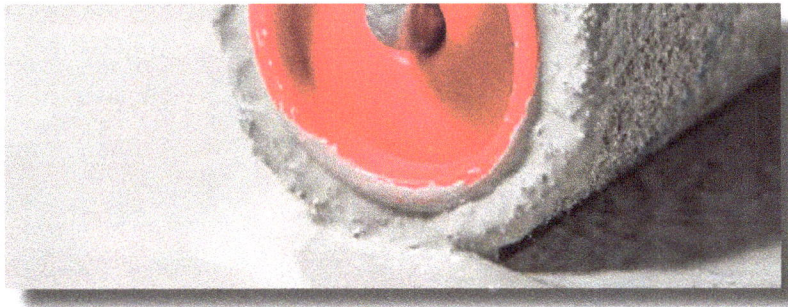

Polyisocyanurate roof insulation is the most used type of insulation in the industry today. It can also be used as a cover board. It is lightweight, has a high R-value, has good fire-resistance properties and is versatile for most roof applications. An example is shown in Figure 1.8.1. The top layer, or "facer," of polyisocyanurate insulation is increasingly being composed of coated glass to overcome the limitations of paper facers that have been historically used.

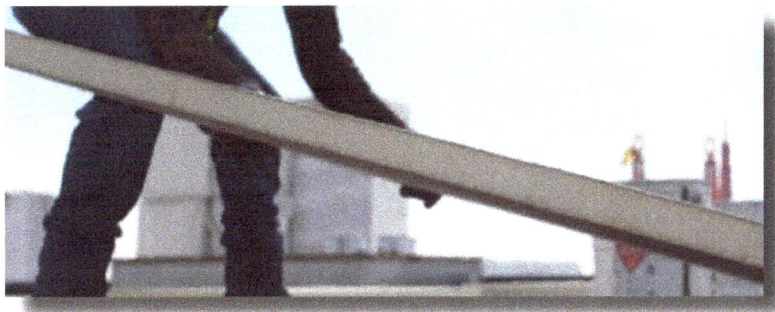

Extruded polystyrene (XPS) and expanded polystyrene (EPS) are two types of rigid foam insulation, as shown in Figures 1.8.2 and 1.8.3. XPS is lightweight and possesses high compressive strength with a moderate R-value. It does not, however, have good fire-resistant properties. Similarly, EPS is also lightweight, with moderate R-values. Its limitations include poor fire resistance properties, compressive strength and water absorption.

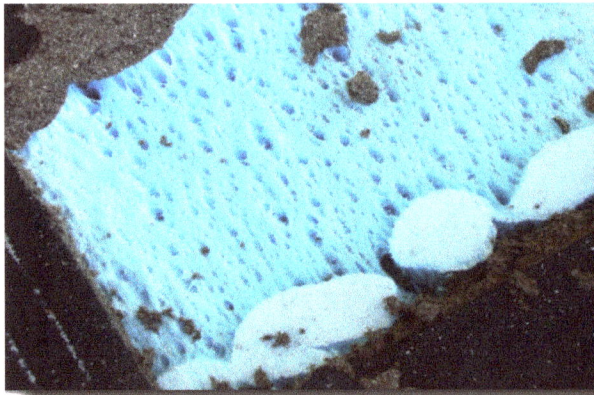

Figure 1.8.2 – XPS Insulation

Figure 1.8.3 – EPS Insulation

Spray polyurethane foam (SPF) is a type of insulation that is sprayed onto the roof. It forms a seamless, monolithic layer that reduces air leakage and thermal bridging. SPF is frequently used in warmer climates that don't experience snow or hail. It is, however, difficult to install and requires specialized equipment and training.

A cover board acts as a thin protective layer within the roof system, adding strength and long-term resiliency. Cover boards protect against hail, foot traffic and other types of impact to the roof's surface. They are typically installed directly below the waterproofing membrane and above the thermal insulation. Some other common materials used as coverboards are shown in Figure 1.8.4.

PERLITE

Gypsum

CEMENT

Wood Fiber

HD Polyiso

Mineral Wool

Asphaltic

Figure 1.8.4 – Cover Boards

13

Accessories on a roof include such items as boots for pipe expansion joints and materials for corner installations. Fasteners play a critical role in how a roof is attached to the deck and resist wind uplift forces. Low-sloped roofs are generally attached by mechanically fastening the systems to the deck, or adhering it. Fasteners come in a variety of shapes and sizes, including nails, screws and bolts. Adhesives may be solvent-based or water-based and can be either liquid-applied or sprayed. Adhered systems prevent flutter and thereby reduce disturbance to the building occupants.

The adhesive market is growing quickly. In 2016, 48% of all low-sloped roofs were adhered, and by 2018, adhesives accounted for 57% of roof attachment.

There are five primary types of adhered systems:

Solvents and adhesives with low volatile organic compound (VOC) content (Figure 1.10.1) are traditional types of adhesives that have been used for years. They perform well in high winds, don't require expensive equipment and have low maintenance costs. They can be applied to a variety of substrates and exhibit good resistance to moisture and weathering.

Foam adhesives (Figure 1.10.2) are applied in liquid form. They then expand and harden into foam. They are often used with single-ply membranes such as TPO or PVC. They exhibit good waterproofing properties and can be used to seal around penetrations.

Figure 1.10.2 – Foam Adhesives

Figure 1.10.3 – Self-adhered Membranes

Self-adhered membranes (Figure 1.10.3) can be applied directly to a roof deck or substrate. They perform like peel-and-stick (self-adhering) membranes. They are easy to install and don't emit any odors. A cold temperature – between 20 and 40F – is required for the primer.

Canister adhesives (Figure 1.10.4) are one of the fastest-growing products in the market. The adhesive is sprayable and comes in pressurized canisters. It is reported to be 30% faster to install than roll-on adhesives. It has minimal flash time and can be used for cold-weather applications with temperatures as low as 25F.

Figure 1.10.4 – Canister Adhesives

CHAPTER

02

Building Defensive Spacing

Standards can be defined as general guidelines for use as a rule or basis of comparison in measuring or judging capacity, content, extent, and value of quality. Codes, on the other hand, are a set of principles or a body of law. Codes reference standards, which indirectly makes standards a part of the law.

Following are a few of the organizations involved in developing roofing-related standards:

- AAMA – American Architectural Manufacturers Association
- ASCE – American Society of Civil Engineers
- ASTM – American Society for Testing and Materials
- CSA – Canadian Standards Association
- CSSB – Cedar Shake and Shingle Bureau
- FM – Factory Mutual
- SPRI – Single Ply Roofing Industry
- UL – Underwriters Laboratory
- WDMA – Window and Door Manufacturers Association

There are four types of standards:

1. Test method (e.g., ASTM E108)
2. Specification and product standard (e.g., ASTM D6878)
3. Practice (e.g., ASTM D7186)
4. Guide (e.g., ASTM D6630)

Guides are not enforceable. Most standards applied in the roofing industry are developed by ASTM, and ASTM publishes more than 12,000 standards annually.

According to the International Building Code (2018), the purpose of the code is "to establish the minimum requirements to provide a reasonable level of safety, public health and general welfare through structural strength, means of egressing facilities, stability, sanitation, adequate light and ventilation, energy conservation, and safety to life from fire, explosion and other hazards, and to provide a reasonable level of safety to firefighters and emergency responders."

From the early 1900s to 1999, there were three model code organizations in the U.S. Most of these organizations began after major fires in metropolitan areas. The three are:

1. Building Officials and Code Administrators International (BOCA), which generally served the Northeast and Midwest regions of the U.S.
2. Southern Building Code Congress (SBCC), which generally served the Southeast regions of the U.S.
3. International Conference of Building Officials – Uniform Building Code (UBC), which generally served the regions west of the Mississippi River

Beginning in 1999, these three organizations – over the course of 6 years – developed and consolidated into the International Code Council (ICC). Referred to as "I-codes," the codes developed by ICC are considered to be model codes that are updated and published on a three-year cycle. Model codes serve s the technical basis for state or local code adoption. Codes are adopted at the state level and enforced at the local level. The code provides minimum legal requirements for building construction and operation. They are enforced by the "authority having jurisdiction (AHJ). Code violations often form the basis for construction claims-related litigation.

The code official's authority over code enforcement ends at the end of the project. An attorney's authority, however, ends depending on the statute of limitations in each state.

A code violation is considered to be evidence of negligence in most states and can cost the contractor a lot of time and money.

## 2.2 The Contractor's Responsibility

Article 3 of AIA General Conditions (published by the American Institute of Architects) covers the contractor's responsibility: "The Contractor is not required to ascertain that the Contract Documents are in accordance with applicable laws, statutes, ordinance, codes, rules and regulations, or lawful orders of public authorities, but the Contractor shall promptly report to the Architect any nonconformity discovered by and made known to the Contractor as a request for information in such a form as the Architect may require."

Article 3 also states that if the contractor fails to perform the obligations stated above, the contractor will have to pay the costs and damages incurred by the owner. If the contractor does perform his obligations, then the contractor is not liable to the owner or architect for any damages. Therefore, a roofing contractor needs to have a good knowledge of state and local codes pertaining to roofing and bring code violations to the attention of the owner and/or architect. Otherwise, the contractor will be held financially responsible.

## 2.3 Building Codes and Standards

Standards are developed by many different organizations and provide established rules of practice or design requirements. They are not laws. Neither are I-codes or the "model codes" developed and published every three years by the International Code Council. These model codes are consensus-based standards that can be adopted by an Authority Having Jurisdiction (AHJ) and amended into law. Once amended and adopted into law, the code is now the building code in effect in that jurisdiction and must be followed as a body of laws.

Roofing codes are part of all 14 of the 2018 complete collection of code books except for one one – ISPSC. I-codes commonly applicable to roofing include:

- IBC – International Building Code
- IRC – International Residential Code
- IEBC – International Existing Building Code
- IECC – International Energy Conservation Code
- IFC – International Fire Code
- IPC – International Plumbing Code

The International Building Code, 2018 edition, has roofing-related requirements in different chapters. However, most of the requirements are covered in Chapter 15. Requirements relevant to roofing can be found in the following:

- Chapter 10 – Means of Egress
- Chapter 12 – Interior Environment
- Chapter 13 – Energy Efficiency
- Chapter 15 – Roof Assemblies and Rooftop Structures
- Chapter 16 – Structural Design
- Chapter 20 – Aluminum
- Chapter 22 – Steel
- Chapter 24 – Glass and Glazing
- Chapter 26 – Plastic

Figure 2.3.1 shows the adoption and enforcement of the International Residential Code in the

U.S. It is interesting to note that contrary to popular belief, many states do not adopt a statewide code, and many states have a statewide code but do not effectively enforce it. Figure 2.3.2 shows the adoption and enforcement of the International Building Code in the U.S. Many more states adopted the Building Code compared to the Residential Code. Knowing this is helpful to insurance companies because the risk will depend on whether a structure is built according to the code. A structure in a geographical location prone to storms and not built up to code is more vulnerable to damage.

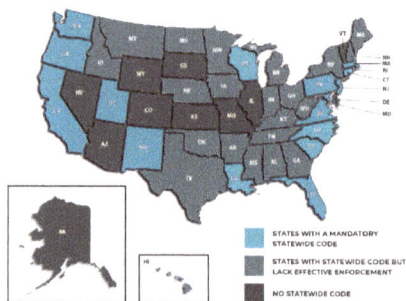

Figure 2.3.1–Adoption and Enforcement of IRC (as of 2023)

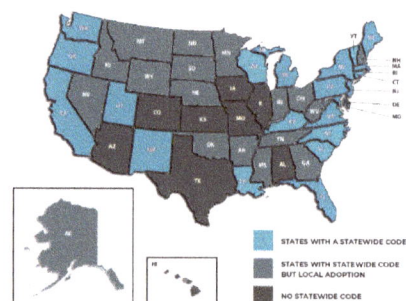

Figure 2.3.2–Adoption and Enforcement of IBC (as of 2023)

Significant roofing requirements pertain to wind resistance, fire classification, installation requirements, prescriptive requirements and reroofing. Chapter 15 covers roof assemblies and rooftop structures.

Codes are written or updated during code development cycles. Multiple committees help review, write and update codes. Bars indicate a revision was made to the code; an arrow indicates a portion of the code was deleted.

Section 1504 from Chapter 15 of the IBC covers wind load performance requirements.

Subsection 1504.1 states that roof decks and roof coverings shall be designed for wind

loads in accordance with Chapter 16, which references ASCE 7-16. This is a standard form

of nomenclature; there are often standard requirements mentioned within codes.

For non-ballasted roofs, Subsection 1504.3 states that the wind load on the roof covering

shall be permitted to be determined using allowable stress design (ASD) and that

mechanically fastened or adhered roofs must be designed to resist wind load pressures.


For asphalt shingles, Subsection 1504.1.1 states that singles shall be wind-tested with

ASTM D7158. Figure 2.4.1 shows a table with classification requirements for the maximum

basic wind speed of the region. Notice that ASTM D3161 assigns an A classification for

the lowest level. This is why it is crucial to have a good understanding of how different

standards work.

TABLE 1504.1.1
CLASSIFICATION OF STEEP SLOPE ROOF SHINGLES TESTED IN ACCORDANCE WITH ASTM D316 OR D71581

| MAXIMUM BASIC WIND SPEED, V, FROM FIGURES 1609.3(1)-(8) OR ASCE 7 (mph) | MAXIMUM ALLOWABLE STRESS DESIGN WIND SPEED, $V_{asd}$ FROM TABLE 1609.3.1 (mph) | ASTM D7158[a] CLASSIFICATION | ASTM D3161 CLASSIFICATION |
|---|---|---|---|
| 110 | 85 | D, G or H | A, D or F |
| 116 | 90 | D, G or H | A, D or F |
| 129 | 100 | G or H | A, D or F |
| 142 | 110 | G or H | F |
| 155 | 120 | G or H | F |
| 168 | 130 | H | F |
| 181 | 140 | H | F |
| 194 | 150 | H | F |

For SI: 1 foot = 304.8 mm; 1 mph = 0.447 m/s.

a. The standard calculations contained in ASTM D7158 assume Exposure Category B or C and building height of 60 feet or less. Additional calculations are required for conditions outside of these assumptions.

Section 1505 of the IBC covers fire resistance requirements. Roof assemblies are classified as Class A, B or C for fire resistance and shall be tested according to ASTM E108 or UL 790 standards. Class A is the highest classification of roof assemblies that are highly effective against fire exposure. Some exceptions to Class A fire testing include coverings with brick/masonry, an exposed concrete roof deck or slate and clay roof tiles.

Figure 2.5.1 shows a table of Minimum Roof Covering Classification for Types of Construction. This table provides the type of roof classification needed for different building types. Notice there is no requirement for Class A roofs. Building Type IA has more stringent fire resistance requirements when compared to Type VB.

**TABLE 1505.1[a, b]**
**MINIMUM ROOF COVERING CLASSIFICATION**
**FOR TYPES OF CONSTRUCTION**

| IA | IB | IIA | IIB | IIIA | IIIB | IV | VA | VB |
|----|----|-----|-----|------|------|----|----|----|
| B  | B  | B   | C[c] | B   | C[c] | B  | B  | C[c] |

For SI: 1 foot = 304.8 mm, 1 square foot = 0.0929 m².

a. Unless otherwise required in accordance with the *International Wildland-Urban Interface Code* or due to the location of the building within a fire district in accordance with Appendix D.

b. Nonclassified roof coverings shall be permitted on buildings of Group R-3 and Group U occupancies, where there is a minimum fire-separation distance of 6 feet measured from the leading edge of the roof.

c. Buildings that are not more than two stories above grade plane and having not more than 6,000 square feet of projected roof area and where there is a minimum 10-foot fire-separation distance from the leading edge of the roof to a lot line on all sides of the building, except for street fronts or public ways, shall be permitted to have roofs of No. 1 cedar or redwood shakes and No. 1 shingles constructed in accordance with Section 1505.7.

Figure 2.5.1- Roof Covering Classification for Types of Construction

Rooftop equipment needs to comply with the requirements of the roof system. For example, photovoltaic panels or vegetative systems placed on the roof become part of the roof and must adhere to roof requirements. This is covered in Sections 1505.9 and 1505.10, as shown in Figure 2.6.1.

**[BF] 1505.9 Rooftop mounted photovoltaic panel systems.** Rooftop rack-mounted *photovoltaic panel systems* shall be tested, *listed* and identified with a fire classification in accordance with UL 1703 and UL 2703. The fire classification shall comply with Table 1505.1 based on the type of construction of the building.

**[BF] 1505.10 Roof gardens and landscaped roofs.** Roof gardens and landscaped roofs shall comply with Section 1505.1 and 1507.16 and shall be installed in accordance with ANSI/SPRI VF-1.

Section 1506 covers material requirements, as shown in Figure 2.6.2. The manufacturer's installation instructions also become part of the code.

## SECTION 1506
## MATERIALS

**1506.1 Scope.** The requirements set forth in this section shall apply to the application of roof-covering materials specified herein. Roof coverings shall be applied in accordance with this chapter and the manufacturer's installation instructions. Installation of roof coverings shall comply with the applicable provisions of Section 1507.

## 2.7 Prescriptive Requirements for Roof Assemblies

Section 1507 covers prescriptive requirements for roof coverings. The code explicitly mentions what needs to be done. The code includes prescriptive requirements for various roof systems, such as asphalt shingles, modified bitumen roofing, metal panels and liquid-applied roofing. It is important to look at both manufacturers' requirements and the prescriptive requirements for roof systems. An example of prescriptive requirements for single-ply roofing is shown in Figure 2.7.1. The table in Figure 2.7.2 shows standards for roof insulation.

**1507.13 Thermoplastic single-ply roofing.** The installation of thermoplastic single-ply roofing shall comply with the provisions of this section.

**1507.13.1 Slope.** Thermoplastic single-ply membrane roofs shall have a design slope of not less than one-fourth unit vertical in 12 units horizontal (2-percent slope).

**1507.13.2 Material standards.** Thermoplastic single-ply roof coverings shall comply with ASTM D4434, ASTM D6754 or ASTM D6878.

**1507.13.3 Ballasted thermoplastic low-slope roofs.** Ballasted thermoplastic low-slope roofs (roof slope < 2:12) shall be installed in accordance with this section and Section 1504.4. Stone used as ballast shall comply with ASTM D448 or ASTM D7655.

Figure 2.7.1–Prescriptive Requirements for Single-ply Roofing

**[BF] TABLE 1508.2**
**MATERIAL STANDARDS FOR ROOF INSULATION**

| | |
|---|---|
| Cellular glass board | ASTM C552 |
| Composite boards | ASTM C1289, Type III, IV, V or VII |
| Expanded polystyrene | ASTM C578 |
| Extruded polystyrene | ASTM C578 |
| Fiber-reinforced gypsum board | ASTM C1278 |
| Glass-faced gypsum board | ASTM C1177 |
| High-density polyisocyanurate board | ASTM C1289, Type II, Class 4 |
| Mineral fiber insulation board | ASTM C726 |
| Perlite board | ASTM C728 |
| Polyisocyanurate board | ASTM C1289, Type I or II |
| Wood fiberboard | ASTM C208, Type II |

Figure 2.7.2–Standards for Roof Insulation

Section 1510 covers rooftop structures. Common rooftop structures include penthouses, tanks, cooling towers, domes, mechanical equipment screens and photovoltaic panels.

The code primarily pertains to new construction, but 70-75% of all roofing work is done on existing buildings. Roofing done on an existing building is called re-roofing. Section 1511 covers reroofing. It states that all requirements of new roofing also apply to reroofing. Roofs can be recovered by installing a new roof covering over an existing roof system, provided a set of requirements is met, as outlined in Figure 2.8.1. Exceptions to roof recovery include roofs with clay, slate, cement or asbestos-cement tiles. Other exceptions are when the roof is water-soaked or deteriorated and when the roof has two or more applications of any type of roof covering. It is very important that the language of codes is clear and interpreted the same way by all parties.

> **1511.3 Roof replacement.** *Roof replacement* shall include the removal of all existing layers of roof coverings down to the roof deck.
>
> > **Exception:** Where the existing roof assembly includes an ice barrier membrane that is adhered to the roof deck, the existing ice barrier membrane shall be permitted to remain in place and covered with an additional layer of ice barrier membrane in accordance with Section 1507.
>
> **1511.3.1 Roof recover.** The installation of a new roof covering over an existing roof covering shall be permitted where any of the following conditions occur:
>
> 1. Where the new roof covering is installed in accordance with the roof covering manufacturer's approved instructions.
> 2. Complete and separate roofing systems, such as standing-seam metal roof panel systems, that are designed to transmit the roof loads directly to the building's structural system and that do not rely on existing roofs and roof coverings for support, shall not require the removal of existing roof coverings.
> 3. Metal panel, metal shingle and concrete and clay tile roof coverings shall be permitted to be installed over existing wood shake roofs when applied in accordance with Section 1511.4.

The IRC covers codes for residential construction. Residential construction comprises one- and two-family dwellings and townhouses no more than three stories in height. Roofing-related requirements for residential construction come under Chapter 8: Roofing/Ceiling Construction and Chapter 9: Roof Assemblies. Chapter 9 closes mirrors Chapter 15 requirements of the International Building Code. The only exception is that the IRC only requires fire-classified roof assemblies where required by local ordinance or where houses are less than three feet from the property line.

The International Existing Building Code pertains to code requirements for existing buildings. Alterations on existing buildings are classified into levels 1, 2 and 3. Roofing comes under Level 1. Alteration Level 1 is considered work on interior finishes, interior floors, window replacement and trim replacement. Section 705 in this book mirrors Section 1511 of the IBC.

The International Energy Conservation Code (IECC) pertains to commercial buildings. Energy code roof requirements cover R-Value, roof reflectivity and air retarders. R-value requirements are broken down by U.S. climate zone3.10.1). R-Values have increased significantly from the 2003 edition to the 2018 edition. For most reroofing work, the insulation must be upgraded to meet the current R-value requirement.

Some definitions to keep in mind include the following:

- Solar reflectance is the fraction of solar flux reflected by a surface expressed within the range of 0.00 and 1.00.

- Thermal emittance is the ratio of radiant heat flux emitted by a surface to that emitted by a black body radiator at the same temperature, expressed within a range of 0.00 to 1.00.
- The Solar Reflectance Index is the relative steady-state surface temperature of a surface with respect to the standard white (SRI=100) and standard black (SRI=0) under standard solar and ambient conditions.

IECC also has codes for air barriers. Test methods include whole building, assembly and material. Certain roof system types, such as built-up, modified bitumen, adhered single-ply and spray polyurethane foam, are deemed to comply with air barrier requirements. An air barrier is not required in reroofing projects until there is also recladding.

## 2.11 Plumbing Code

The International Plumbing Code (IPC) applies to all plumbing systems except those applicable to IRC. Roofing-related plumbing codes come under Chapter 11: Storm Drainage. The IBCC has an entire section (1502) on roof drainage. This shifts the responsibility for roof inspection from the plumbing code officially to the building code official.

## 2.12 Fire Code

The International Fire Code (IFC) covers fire safety requirements. A building code official's authority ends once the Certificate of Occupancy has been delivered to the owner. However, a fire code official can access the building at any time. A fire code official primarily has authority over operations. Roofing-related provisions of the IFC include:

- Section 303: Asphalt Kettles
- Section 317: Rooftop Gardens
- Section 1204: Solar Photovoltaic Systems
- Section 3317: Safeguarding Roofing Operations.

For other materials, design, methods of construction and equipment, some of these books have a section mostly around Section 104.11, which allows the code official the authority to grant permission on code waivers for other materials, provided they are asked beforehand. Building a good relationship with code officials and clearly communicating waiver requirements are vital so that the officials are more likely to sign off on inspections.

## 2.13 Reference Standards

Building codes reference standards. Common types of standards include test method (fire test, wind test, hail test, etc.), specification or product standard (products need to meet certain standards), practice (best practice for activities), and guide (how to perform certain activities efficiently). The American Society for Testing and Materials (ASTM), Factory Mutual (FM) and

Underwriters Laboratories (UL) are important groups and organizations that publish standards.

UL is a test standards and certification company, but they do not offer insurance. They test, certify, and inspect materials, products, and systems to provide classifications.

Common UL standards related to roofing are:
- UL 790: Standard Test Methods for Fire Tests of Roof Coverings
- UL 2218: Impact Resistance of Prepared Roof Covering Materials
- UL 1897: Uplift Tests for Roof Covering Systems
- UL 580: Standard for Tests for Uplift Resistance of Roof Assemblies

FM Global is part of a suite of companies that include FM Approvals, AFM, FM Global Cargo, Mutual Boiler Re, and Emergency Response Consultants. FM Approval is another company that creates test standards and certifies products/services. They, however, do also provide insurance and consultancy services on a project.

FM Approvals cover testing standards and certifications for fire protection and detection equipment, building materials, roofing assemblies, and wall assemblies.

ASTM International also provides standards. They have a vast group of technical committees, some of which include:

- D08 – Roofing and waterproofing

- E06 – Performance of Buildings (Metal roofing, EIFS, Glazing)

- D20 – Plastics

- C27 – Precast concrete

- F08 – Sports equipment, playing surfaces, and facilities

- F15 – Consumer products (Toys, auditorium seating, brooms and mops, furniture, lighters, etc.).

The ASTM website has a complete list of standards and their information.

# 03

The Roofing Industry and
the Environment

With content provided by
Heather Hardy-Forino
GAF Corp

The 1987 report Our Common Future includes a classic definition of sustainability: to "meet the needs of the present without compromising the ability of future generations to meet their own needs." Another common definition looks at sustainability as an intersection of society (human and well-being), the environment (environmental resource abundance) and the economy (economic vitality and prosperity), achieving a balance of the three (Figure 3.1.1).

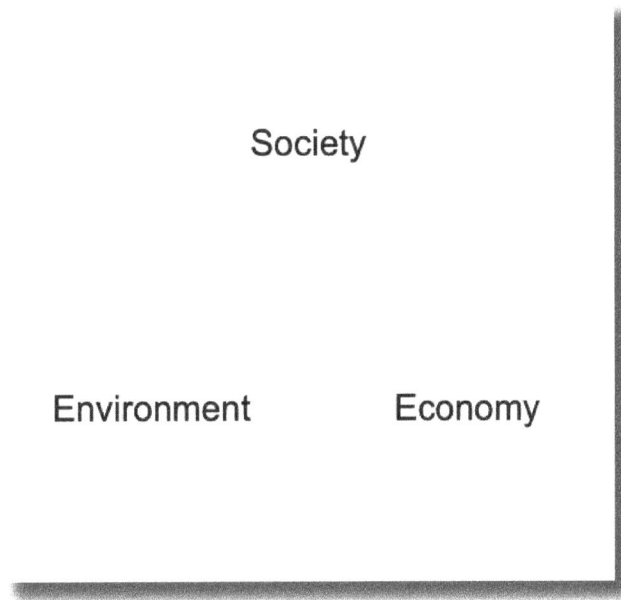

**Society**

**Environment**          **Economy**

Figure 3.1.1: Sustainability

Sustainability in roofing primarily involves designing, installing and maintaining roofs in a way that minimizes their impact on the environment and increases their longevity and energy efficiency. One way that roofing contributes to sustainability is through the use of sustainable materials. Another way is by maximizing energy efficiency because an energy-efficient roof system enables the building to use less energy for heating or cooling. An example is a "cool roof" that reflects part of the sun's radiation. Proper design and installation will ensure that the roof is well-insulated and ventilated, which leads to reduced energy loss and moisture buildup. Regular maintenance will promote longevity and reduce the need for replacement.

Roof demolition can create a large amount of waste, including some hazardous materials that may contain asbestos.

Sustainable practices during demolition include salvation/reuse of materials and proper handling and disposing of toxic materials. "Green" or vegetative roof systems are increasingly being used to reduce heat absorption and provide insulation.

## 3.2 Rating Systems

There are three rating systems for green – or environmentally friendly – buildings. They are:

- LEED (Leadership in Energy and Environmental Design), developed by the U.S. Green Building Council (USGBC).
- Living Building Challenge
- Green Globes

LEED is the world's most widely used green building rating system, and its certification is a globally recognized symbol of sustainability achievement and leadership. LEED has a large number of categories that can earn credits, and every project must earn credits to achieve certification. The four types of certification are LEED Certified (40-49 credits), LEED Silver (50-59 credits), LEED Gold (60-79 credits) and LEED Platinum (80 credits or more).

Typical roofing contributions to LEED (Figure 3.2.1) include:

- Sustainable sites. Roofing materials play a huge part in heat island reduction. An example is a "cool" – or reflective – roof.
- Materials and resources. Environmental Product Declarations (EPDs), sourcing of raw materials, material ingredients and construction/demolition waste planning and management.

| System and Version | | LEED®[1] v4 FOR BUILDING DESIGN AND CONSTRUCTION: NEW CONSTRUCTION | | | | | | |
|---|---|---|---|---|---|---|---|---|
| Credit Category | Sustainable Sites | Energy and Atmosphere | | Materials and Resources | | | | Indoor Environmental Quality |
| | SSc5 | EAp2/EAc2 | EAc5 | MRc2 | MRc3 | MRc4 | MRp2/MRc5 | EQc2 |
| Credit Description | Heat Island Reduction | Min/Optimize Energy Performance | Renewable Energy Production | BPDO – Environmental Product Declarations (EPDs) | BPDO – Sourcing of Raw Materials (e.g., recycled content) | BPDO – Material Ingredients (e.g., HPDs) | Construction & Demo Waste Planning/ Management | Low-Emitting Materials |
| | Opt 1 & 2 | | | Option 1 | Option 2 | Option 1 | Opt 1 & 2 | Opt 1 & 2 |

The Living Building Challenge is a building certification program developed by the International

Living Future Institute in 2006. It is centered around the aspirational question: "What if every single act of design and construction made the world a better place?" It comprises seven performance categories, or "petals," that include place, water, energy, health and happiness, materials, equity and beauty. Figure 3.2.2 shows where negative impacts on the environment keep reducing as codes, green requirements and high-performance requirements are met. Positive environmental impacts are created once sustainability requirements are met. LEED lies in between green and high-performance requirements.

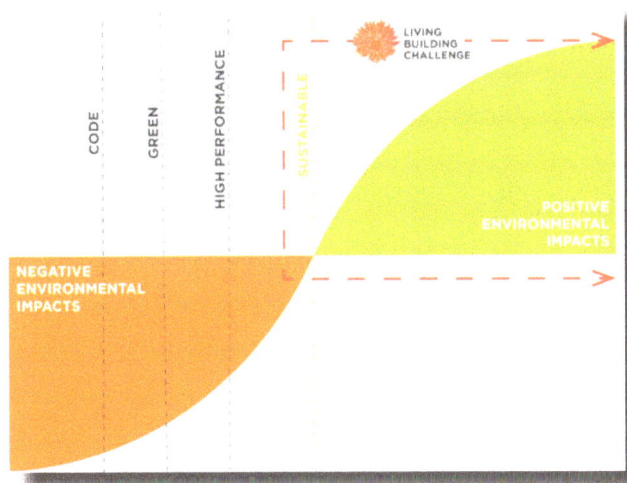

Typical roofing contributions to the Living Building Challenge lie in the Health and Happiness and Materials petals. The Health and Happiness petal focuses on requirements that promote the physical and emotional health and well-being of building occupants. As it applies to roofing, this petal includes requirements for daylighting, indoor air quality, noise reduction, etc. The Materials petal focuses on requirements that promote the use of sustainable and responsible materials. For roofing, this includes materials not on the "red list" developed by the International Union for Conservation of Nature, "declare labels" developed by the International Living Future Institute, material sourcing and net positive waste.

Green Globes is very similar to LEED. Its evaluation tool is easy to use and covers new construction as well as maintenance and repair work. The Green Globes program was developed by the Green Building Initiative (GBI), which started in Canada and expanded to the U.S. in 2004. Typical roof contributions to Green Globes include site (heat island effect) and energy (thermal resistance, on-site renewable energy).

Some third-party sources provide all the information needed for environmental ratings of products and services. An example is ecoScorecard. Figure 3.2.3 shows an ecoScorecard for TPO roof membrane. It provides environmental certifications, including the declare label (containing information about chemical ingredients and materials used in the product), EPD documentation, Health Product Declaration, and so on. It provides environmental characteristics such as the level of volatile organic compounds (VOCs), emissivity and reflectivity of the product.

Category: Roof Products
Category: Roof Products

# GAF

EverGuard® TPO (white) 45mil, 60mil, 80mil

GAF EverGuard® TPO(white) 45 mil, 60 mil & 80 mil thermoplastic polyolefin single-ply membrane systems have been engineered to provide premium performance at a very cost-effective price. Strong, flexible, durable EverGuard® TPO 45 mil, 60 mil & 80 mil membrane is suitable for use in all types of single-ply systems: Mechanically Attached, Fully Adhered, and Ballast Applied.

## Rating Systems, Standards, & Frameworks

Check credits for mindful Materials, LEED, WELL, Living Building Challenge, and other ratings systems, standards, and frameworks. To access sign-in or register for free.

## Sustainability Highlights

### Mindful Materials

✓ Human Health
✓ Ecosystem Health
✓ Climate Health

### Certifications & Ecolabels

✓ HPD
✓ DECLARE® - Red List Free
✓ EPD
✓ GreenCircle

### Ratings Systems & Standards

✓ LEED
✓ Living Building Challenge
✓ WELL

### LEED Credits

✓ Materials and Resources
✓ Green infrastructure & buildings
✓ Sustainable Sites

### Circularity

✓ 100% Recyclable

### Carbon

✓ LCA

# Details

📍 **Is this your project location? Edit the location at the top of the page for plant specific information**

Postal Code: 29634

## Certifications and Standards

Environmental Product Declaration (EPD)

Certified by
valid 2019-12-11 to 2024-12-11

GreenCircle Certificate

Certified by
valid 2024-08-01 to 2025-07-31

Health Product Declaration (HPD)

Sustainable Minds, as shown in Figure 3.2.4, is another third-party source that offers

a great tool for designers to assess the sustainability of products and operations. The

environmental and social impact of products can be assessed through the product's

lifecycle.

## Rating systems

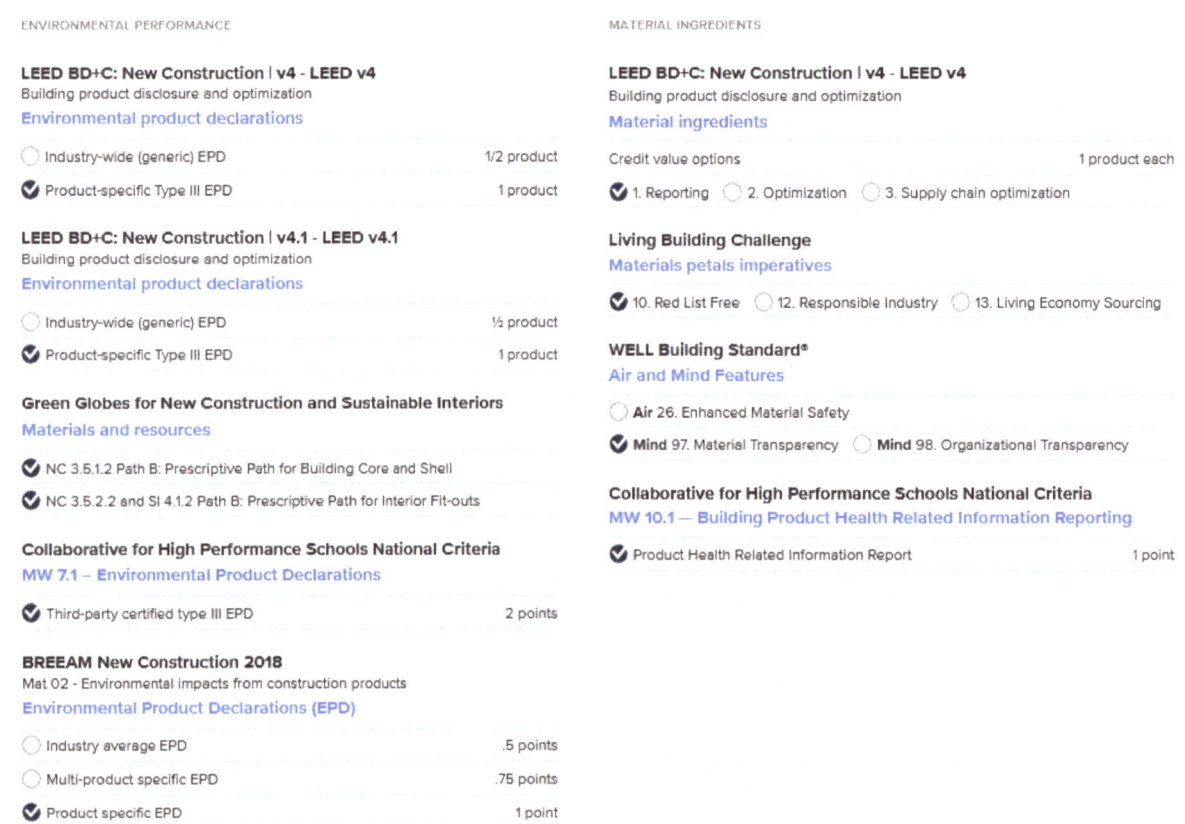

**ENVIRONMENTAL PERFORMANCE**

**LEED BD+C: New Construction | v4 - LEED v4**
Building product disclosure and optimization
Environmental product declarations

| | |
|---|---|
| ○ Industry-wide (generic) EPD | 1/2 product |
| ✓ Product-specific Type III EPD | 1 product |

**LEED BD+C: New Construction | v4.1 - LEED v4.1**
Building product disclosure and optimization
Environmental product declarations

| | |
|---|---|
| ○ Industry-wide (generic) EPD | ½ product |
| ✓ Product-specific Type III EPD | 1 product |

**Green Globes for New Construction and Sustainable Interiors**
Materials and resources

✓ NC 3.5.1.2 Path B: Prescriptive Path for Building Core and Shell

✓ NC 3.5.2.2 and SI 4.1.2 Path B: Prescriptive Path for Interior Fit-outs

**Collaborative for High Performance Schools National Criteria**
MW 7.1 – Environmental Product Declarations

| | |
|---|---|
| ✓ Third-party certified type III EPD | 2 points |

**BREEAM New Construction 2018**
Mat 02 - Environmental impacts from construction products
Environmental Product Declarations (EPD)

| | |
|---|---|
| ○ Industry average EPD | .5 points |
| ○ Multi-product specific EPD | .75 points |
| ✓ Product specific EPD | 1 point |

**MATERIAL INGREDIENTS**

**LEED BD+C: New Construction | v4 - LEED v4**
Building product disclosure and optimization
Material ingredients

| | |
|---|---|
| Credit value options | 1 product each |

✓ 1. Reporting ○ 2. Optimization ○ 3. Supply chain optimization

**Living Building Challenge**
Materials petals imperatives

✓ 10. Red List Free ○ 12. Responsible Industry ○ 13. Living Economy Sourcing

**WELL Building Standard®**
Air and Mind Features

○ **Air** 26. Enhanced Material Safety

✓ **Mind** 97. Material Transparency ○ **Mind** 98. Organizational Transparency

**Collaborative for High Performance Schools National Criteria**
MW 10.1 — Building Product Health Related Information Reporting

| | |
|---|---|
| ✓ Product Health Related Information Report | 1 point |

Figure 3.2.4 – Sustainable Minds

Declare labels provide a list of all ingredients used in the product and indicates whether the

product's materials are red-list free for compliance with rating terms. The red list contains

materials that cause harm to humans or the environment.

The Environmental Product Declaration (EPD) is a comprehensive report that. Provides third-party verified information about the environmental impact of products throughout their lifecycle, from the extractor of the raw materials to their disposal after use. An example is shown in Figure 3.2.5.

## Environmental Product Declaration

GAF EverGuard® TPO Membrane
Single Ply Roofing Membrane (TPO)

**GAF NSF**

According to
ISO 14025, ISO 14044,
and ISO 21930:2017

This declaration is an environmental product declaration (EPD) in accordance with ISO 14025 and ISO 21930-2017. EPDs rely on Life Cycle Assessment (LCA) to provide information on a number of environmental impacts of products over their life cycle. Exclusions: EPDs do not indicate that any environmental or social performance benchmarks are met, and there may be impacts that they do not encompass. LCAs do not typically address the site-specific environmental impacts of raw material extraction, nor are they meant to assess human health toxicity. EPDs can complement but cannot replace tools and certifications that are designed to address these impacts and/or set performance thresholds – e.g., Type 1 certifications, health assessments and declarations, environmental impact assessments, etc. Accuracy of Results: EPDs regularly rely on estimations of impacts, and the level of accuracy in estimation of effect differs for any particular product line and reported impact. Comparability: EPDs are not comparative assertions and are either not comparable or have limited comparability when they cover different life cycle stages, are based on different product category rules or are missing relevant environmental impacts. EPDs from different programs may not be comparable.

| EPD PROGRAM AND PROGRAM OPERATOR NAME, ADDRESS, LOGO, AND WEBSITE | NSF International, 789 N. Dixboro Rd, Ann Arbor, MI 48105,www.nsf.org |
| --- | --- |
| GENERAL PROGRAM INSTRUCTIONS AND VERSION NUMBER | NSF Certification Policies for Environmental Product Declarations (EPD): November 1, 2022 |
| MANUFACTURER NAME AND ADDRESS | GAF<br>1 Campus Drive,<br>Parsippany, NJ 07054 |
| DECLARATION NUMBER | EPD10913 |
| DECLARED PRODUCT & DECLARED UNIT | GAF EverGuard® TPO Membrane<br>Declared Unit = 1 m² |
| REFERENCE PCR AND VERSION NUMBER | NSF International: Product Category Rule for Environmental Product Declarations for Single Ply Roofing Membranes, Version 2. Issued 2019<br>Valid through July 17, 2024 |
| DESCRIPTION OF PRODUCT APPLICATION/USE | Single Ply Roofing Membrane (TPO) |
| PRODUCT RSL DESCRIPTION | N/A |
| MARKETS OF APPLICABILITY | Global |
| DATE OF ISSUE | 12/08/2023 - 12/08/2028 |
| PERIOD OF VALIDITY | 5 Years |
| EPD TYPE | Product Specific |
| DATASET VARIABILITY | N/A |
| EPD SCOPE | Cradle-to-Gate w/options |
| YEAR(S) OF REPORTED PRIMARY DATA | 2021 |
| LCA SOFTWARE & VERSION NUMBER | LCA for Experts v. 10.6<br>GAF EPD Generator Tool Version 1.0 |
| LCI DATABASE(S) & VERSION NUMBER | Sphera database & USLCI v2.0 |
| LCIA METHODOLOGY & VERSION NUMBER | TRACI 2.1; CML 4.1 |
| The sub-category PCR review was conducted by: | |
| This declaration was independently verified in accordance with ISO 14025: 2006. The UL Environment "Part A: Calculation Rules for the Life Cycle Assessment and Requirements on the Project Report." v3.2 (Dec 2018), based on ISO 21930:2017, serves as the core PCR, with additional considerations from CEN Norm EN 15804 (2013) and the USGBC/UL Environment Part A Enhancement (2017)<br>☐ INTERNAL    ☒ EXTERNAL | Jack Gelbig, EcoForm, LLC jgelbig@ecoform.com |
| This life cycle assessment was conducted in accordance with ISO 14044 and the reference PCR by: | Sustainable Solutions Corporation |
| This life cycle assessment was independently verified in accordance with ISO 14044 and the reference PCR by: | Jack Gelbig, EcoForm, LLC jgelbig@ecoform.com |

Environmental declarations from different programs (ISO 14025) may not be comparable.

Comparison of the environmental performance using EPD information shall consider all relevant information modules over the full life cycle of the products within the building.

Apart from the contribution that products and manufacturing make to building sustainability, there are several ways that the industry contributes to sustainability. There are manufacturing plants with plant waste diversion certificates, indicating the amount of waste they divert from landfills to recycling or reuse. Among manufacturers, a sustainable supply chain system called EcoVadis assesses whether purchases are made from responsible vendors. Solar installations for energy generation can be seen at roofing material plants and warehouses. There are recycling programs for roofing waste that work on methods for repurposing waste for future use. Water reclamation and construction waste diversion are other factors that contractors and manufacturers are increasingly assessing.

## 3.3 Resiliency and Roofing

Resilience is defined as the capacity to adapt to changing conditions and to maintain or regain functionality in the face of stress or disturbance. For a roof, resilience means performing through a storm or other severe weather conditions. Resiliency is an essential quality for a roof system because the roof is the "fifth wall" of every building. The roof provides basic shelter while protecting life and the property in the building.

Sustainability and resiliency are often considered two separate concepts, leading to the predominance of sustainability over resiliency. However, they are concepts that intersect, producing trade-offs. An example of this intersection in roofing is that rainwater harvesting could be less resilient but more sustainable while using a greywater system could be both more sustainable and more resilient. Rainwater harvesting, daylighting and solar panels with storage capacity are more sustainable while also potentially making the building more resilient.

Roofs need to be designed to perform under changing conditions. A tougher roof is a more resilient roof. The "toughness" of the roof system will be measured against such factors as wind loads, fire resistance, impact resistance and so on. Designing for wind should follow the simple logic that the uplift resistance of the roof must be greater than wind loads. Wind loads acting on the roof will depend on the following:

- Location of the building

- Building code in effect at the building's location

- Height, length and width of the roof

- Exposure (depending on surrounding buildings, trees, etc.)

- Use and occupancy of the building

- Type of building

If the same type of building were to be placed in two different locations, the wind design for the roofs would be very different. For example, if a two-story office building of the same size were located in both Dallas and Houston, the Dallas roof would need to provide 40 psf at the field of the roof, and the Houston roof would need to provide 59 psf because of higher wind speeds in Houston. This number would be still higher at the roof's corners.

While designing for wind, it is imperative to consider the weakest link in the roof. For example, substrate attachment is the most critical element for preventing failure in a fully adhered roof system. It is always a good practice to design and install roofs not only according to the code, but beyond the code requirements to achieve additional resiliency.

Roof performance during high-wind events can be enhanced using parapet walls, air barriers and non-permeable decks. The weakest link must be identified and removed, or its performance must be enhanced. Opting for full-adhered roof systems instead of mechanically attached roof systems and increasing the cleat thickness of metal edges can also enhance roof performance.

## 3.5 Impact Resistance

Hail activity and frequency in different regions of the country. Roof designs need to account for impact resistance from hail, depending on the location of the building. Impact resistance depends on the roof substrate and how the system has been fastened. A rigid, high-density cover board will significantly contribute toward keeping the roof resistant to hail damage.

Roofs need to be built for toughness upon impact, not just for protection from hail damage. There are four ways to ensure the toughness of the roofing system, as follows:

- No fasteners directly under the membrane
- Rigid cover board
- Fully adhered roof membrane
- Supported base flashings

There are multiple test standards for wind and impact testing, such as those developed by FM and UL. Resources such as RoofNav (from FM) and the UL Product Directory provide information about wind uplift resistance, hail ratings for roof system design and installing a resilient roof system.

The type of installation dictates the resiliency of the roof system—whether it is mechanically attached or fully adhered. Generally, a fully adhered roof system with a cover board and buried fasteners will provide more resiliency but will probably have a higher initial cost.

## 3.6 Daylighting

One of the ways a roof can contribute to the sustainability and resiliency of a building is through daylighting, which will ensure that the building is inhabitable even during a power outage. Natural light can be brought into a building through the roof using skylights or tubular daylight devices, as shown in Figure 4.6.1. Daylighting is more than providing skylights. Daylight devices need to be engineered for the roof's size, location and reflectivity.

Providing for skylights will result in the loss of insulation on the rooftop. However, this can be balanced with the energy savings daylights can bring. The IECC 2015 edition states that "skylight curbs must be insulated to the level of the roofs with insulation entirely above deck or R-5, whichever is less."

## 3.7 Roof Insulation and Resiliency

The purpose of insulation is to control climate differences by keeping cold air in and hot air out—or vice versa, depending on conditions. Insulation is important for environmental control and occupant comfort, and it contributes to energy efficiency – adding to building resiliency.

Cool roofs reflect 80% of the sunlight that falls on them, whereas dark roofs only reflect 20%. Therefore, using cool roofs in warmer climates to reduce heat gain usually makes sense.

The IECC residential provision applies to all residential buildings less than three stories in height. The three common paths for code compliance include ASHRAE 90.1, Prescriptive Path and Performance Path. Contractors most commonly follow the Prescriptive Path due to its straightforward nature. According to the climatic zone of the location, the minimum R-value requirements can be found in the ICC building codes.

However, many other steps can be taken "beyond the code" to enhance roof resiliency. These include:

- Using two layers of insulation with staggered joints, as shown in Figure 3.7.1, to prevent energy loss.

- Reducing or preventing thermal bridging caused by fasteners. Fasteners act as conductors of heat and result in energy loss. There is an estimated 5-12% energy loss through thermal bridging. Fasteners try to normalize exterior and interior temperatures. An example is shown in Figure 3.7.2, where heat conducted by fasteners has created dimples on the snow

Figure 3.7.1 - Staggered Insulation

Figure 3.7.2 - Melting of Snow Over the Fasteners

Most building owners want a roof that, first, is installed and maintained to be resilient. Later, an owner may consider options for energy savings, including using rooftop solar systems. For a building owner – and for a contractor or architect – a large roof area is great real estate. It can be used, for example, to house mechanical equipment. It can also be used to install solar panels to generate power, save energy and increase sustainability. Figure 3.8.1 shows thin-film solar panels adhered to a roof. If done correctly, these can help make a roof very resilient and reduce its dependency on the power grid.

The following are a few things to consider for rooftop solar installation:

- How the roof system is attached, and how the solar panels will be attached

- The age of the roof. There is no point in attaching solar panels on a poorly performing roof that will need replacement within the next ten years.

- The potential for shade. Solar panels need to be placed in areas mostly unshaded to perform effectively.

- The type of roof system that will be used as a substrate and whether it is mechanically attached or adhered.

- The cost of poor decisions

- Access to the rooftop and the underlying roof system

- How the roof will be maintained

- Drainage on the roof

- Other equipment on the roof

- Code and regulatory requirements

Figure 3.8.2 is an example of a shopping center's roof where improper drainage of the roof led to weeds growing in between the solar panels. These weeds cast shadows and affect the overall efficiency of the panels.

Figure 3.8.2–Weeds Growing through Solar Panels

Figure 3.8.3 shows a cold storage facility near a lake. Slightly higher wind speeds entirely damaged the ballasted solar system. Figure 3.8.4 shows a large convention center located 30 miles from the cold storage facility. This building had solar panels affixed to the roof and yet did not suffer any damage due to the roof being well-designed. It is interesting to note how the solar panels have been placed precisely, aligning with the shadow pattern on the roof.

In summary, to ensure roof resiliency, considering wind resistance, impact resistance, daylighting, insulation, surface color and rooftop energy production is essential.

Roof Scheduling

A general contractor (GC) focuses on meeting the desired end date of a project and does not always understand the ramifications of schedule changes. GCs work in the best interests of the overall project and foster cooperation among the various specialty contractors (SCs) involved. A construction project usually has multiple specialty contractors involved.

Specialty contractors focus primarily on their specific scope of work when determining the schedule. They can help the overall project by identifying different approaches to their work. Often, some of the SCs that work later in the project cycle might have to bear the burden of previous delays.

The GC's project scheduler typically issues an agenda for the overall project and meets with stakeholders in advance to discuss milestone events and project expectations. An example of an agenda is shown in Figure 4.1.1. The scheduler's agenda is usually discussed either weekly or biweekly with the entire project team and includes important project information, such as procurement, site work and general details. Not all items are pertinent to every stakeholder, but these meetings help ensure that all parties know what is expected of them and when.

The general contractor manages the overall schedule. The scheduler gets input from the various SCs involved to determine the duration of their scope of work. The schedule has critical dates, the project's overall timeline and key activities. Schedules are typically modified on a two- to three-week basis. Critical activities are color-coded. An example of a six-week schedule for a project is shown in Figure 4.1.2.

Figure 4.1.1–Scheduler Agenda

Figure 4.1.2–Project Schedule

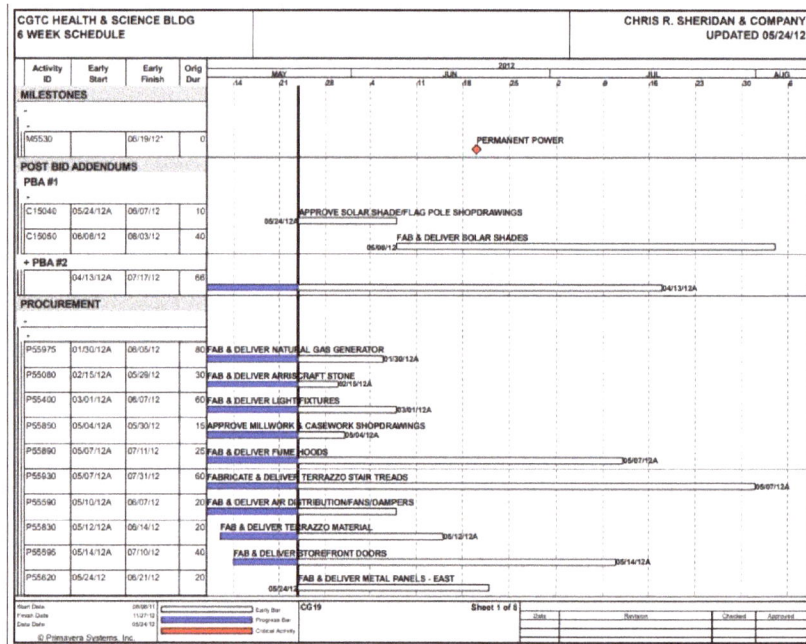

After every meeting, a list of action items is made, as shown in Figure 4.1.3. This list describes the tasks to be done, who is responsible for each task and when it must be completed. This is a great way to ensure stakeholders are held accountable for solving problems and preventing project delays.

| ITEM NO | DESCRIPTION | RESPONSIBILITY | DATE REQUIRED | DATE RESOLVED |
|---------|-------------|----------------|---------------|---------------|
| 13.1 | The flashing detail with the brick and brick ties at the mock up needs to be resolved. | Architect | 3/2/12 | 3/2/12 |
| 13.2 | The duct seal material needs to be approved for LEED. | Hamlin Air Conditioning | Noted | |
| 13.3 | An RFI needs to be submitted about all mounting height information needed for electrical rough ins. | Starr Electric | 3/5/12 | 3/5/12 |
| 13.4 | Delta Fire Protection has requested as letter of certification on the fire line. | Pyles Plumbing | 3/5/12 | 3/30/12 |
| 13.5 | Information is needed on the dimming system.  This affects inwall rough ins. | Architect | 3/7/12 | 3/7/12 |
| 13.6 | The Engineer will be reviewing inwall rough ins.  Plumbing is ready on the 1st floor but electrical still needs to much information.  The inspections need to be coordinated. | Engineer | Noted | |
| 13.6.1 | HESMA and the Engineers have been on site reviewing work in place.  They will be called for a final review just prior to sheetrock. | Engineer | 5/29/12 | |
| 13.6.2 | HESMA will be brought in beginning 5/21/12 to begin reviewing all work in place. | HESMA | 5/21/12 | |

A "look ahead" schedule for a roofing contractor, as shown in Figure 4.1.4, is sometimes used to provide an outlook of upcoming tasks and activities to manage the roofing project internally. This helps with planning and ensuring enough resources are available for the work required. In addition, as shown in Figure 4.1.5, a weekly planner might be used to create a plan for the week, including to-do activities, outstanding activities from previous weeks and goals to be accomplished by the end of the week.

Some considerations while scheduling a roofing project include:

**Temporary roofing**

Some projects require a temporary roof to protect the structure from weather damage and keep it dry for interior construction work. Temporary roofs should be planned so they do not cause delays to the rest of the activities on a project. For steep-slope roofs, synthetic underlayment is often used in peel-and-stick form or fastened, depending on how long it will remain exposed. For low-slope roofs, temporary roofs are often costlier due to the high chance of water seepage. In some cases, the temporary roofing material can also be used as a vapor retarder if required. Temporary roofing can be used efficiently in reroofing projects if appropriately planned.

**Rooftop equipment**

It is common to see multiple pieces of equipment (mechanical, HVAC, etc.) found in the field of the roof. Leaks occur most frequently in transition areas, and special attention has to be given to flashings. If the building has a parapet wall, it needs to be constructed prior to the installation of the roof system since the roof system must wrap up over the parapet wall. The type of exterior façade the building has must also be considered. When installing scuppers, the roofing contractor must ensure that the roof's interior parts are waterproofed. The contractor responsible for the exterior wall must ensure the scupper goes through the wall and that surrounding areas are waterproofed. The type of drainage system used, exterior weather barrier, and perimeter safety are other considerations. This is why discussing the best possible sequencing for a project is essential.

Protecting the roof from potential damage is always an important consideration. Multi-ply roof systems are much more resistant to damage but are relatively expensive to install and are no longer used frequently.

Damage protection is much more challenging for single-ply roof systems while being installed. Regular foot traffic on a newly installed white single-ply roof can cause membrane discoloration and may impact energy efficiency. With multiple crews accessing the roof for other types of work, finding damage and holding the right party responsible is often challenging. However, it is an essential part of the project's management.

## 4.3 Schedule Coordination

Roofing on any type of construction project is always a critical step. A preconstruction conference involving the general contractor and all subcontractors involved with the roof (e.g., mechanical, plumbing and decking contractors) must be arranged. This will help to ensure that the scope of work and the general sequence of work is known to everyone.

Before roofing work can begin, the building structure should be examined to be certain it is ready for the roof to be installed. This includes ensuring that the roof deck is installed correctly and has the proper slope. The roofing contractor must also understand what other contractors will be involved in the project when scheduling the installation. It is always better to have a continuous flow of work.

## 4.4 Creating a roof schedule

Excel or other types of scheduling software are commonly used to create a project schedule. Generally, the type of software selected will depend on the project's size and complexity and the type of software the company uses internally.

To begin creating a schedule, the following information needs to be collected:

- The roof estimate, including quantities of materials - It is critical to know the square footage of the roof, the linear footage of flashings to be installed and so on.

- Crew size - The type and number of people to be assigned to the project need to be finalized so that the time needed for installation can be determined.

- Productivity rate - Every roofing contractor keeps a record of productivity for each crew and crew member. This will enable the contractor to determine how many man-hours are needed to complete various activities on the project.

- Major activities - The contractor must thoroughly understand the scope of work.

- Crew days for each activity - All the information gathered at this stage will help determine how many crew days are required for each activity and for the overall project.

## 4.5 Scheduling Example

Let's consider how to calculate the number of crew days for a roofing project that involves installing 2,500 square feet of TPO membrane with 2.2" of polyisocyanurate roof insulation and also involves 350 feet of wall flashing for a 3' high parapet wall. One crew consisting of a foreman, two mechanics and two laborers/helpers will be used (Figure 4.5.1).

| Roofing and Membrane Flashing Labor | Production Rate | Unit of Measure |
|---|---|---|
| Install TPO Adhered \|Coverboard \| Tapered Iso (Concrete) | 30 SF | Per Man Hour |
| Install TPO Adhered \|Coverboard \| Tapered Iso (Metal Deck) | 25 SF | Per Man Hour |
| TPO Membrane Wall Flashing (1' Height) | 4 LF | Per Man Hour |
| TPO Membrane Wall Flashing (2' Height) | 3.5 LF | Per Man Hour |
| TPO Membrane Wall Flashing (3' Height) | 3 LF | Per Man Hour |
| TPO Membrane Wall Flashing (4+' Height) | 2.5 LF | Per Man Hour |
| TPO Membrane Curb Flashing | 3.25 LF | Per Man Hour |
| TPO Thru Roof Pipe Penetration (Pipe Boot) | 2 EA | Per Man Hour |
| TPO Thru Roof Pipe Penetration (Field Fabricated) | 1 EA | Per Man Hour |
| TPO Drip Edge Membrane "Strip In" | 5 LF | Per Man Hour |
| TPO Roof Drain Flashing | 0.5 EA | Per Man Hour |
| Wood Blocking Install | 10 LF | Per Man Hour |

Figure 4.5.1 - Productivity Rates

According to the productivity rates provided in Figure 5.5.1, man hours (MH) required to install 30 feet of TPO are:

$$\frac{2,500 \text{ sf of TPO}}{30 \text{ sf}} = 83.4 \text{ total MH}$$

Assuming that the five-person crew will work 8 hours daily, we get 40 total Crew Hours (CH). The crew day will be calculated as:

$$\frac{83.4 \text{ MH}}{40 \text{ CH}} = 2.085 \text{ crew days}$$

Similarly, for installing flashing,

$$\frac{350 \text{ feet}}{3 \text{ linear feet}} = 116 \text{ MH}$$

$$\frac{116 \text{ MH}}{40 \text{ CH}} = 3 \text{ crew days}$$

It is important to note that in addition to the major activities of the project, there are additional activities that also need to be accounted for while creating a schedule. These include such things as preparing and getting approval for shop drawings, mobilization and job set-up, handling punch list items and demobilization.

It is also critical to note that all activities listed do not occur completely linearly, i.e., one right after the other. There might be a significant overlap that can be worked into the final schedule.

Figure 4.5.2 shows a roof schedule for replacing the roof from areas A to G on a building. Even though the roofing work begins at the bottom half of the schedule, management activities need to be done before the work begins and need to be accounted for in the top half of the schedule. Notice that many activities have overlaps with the previous and ensuing activities. Understanding the relationships among these activities is essential to determining where the overlaps are.

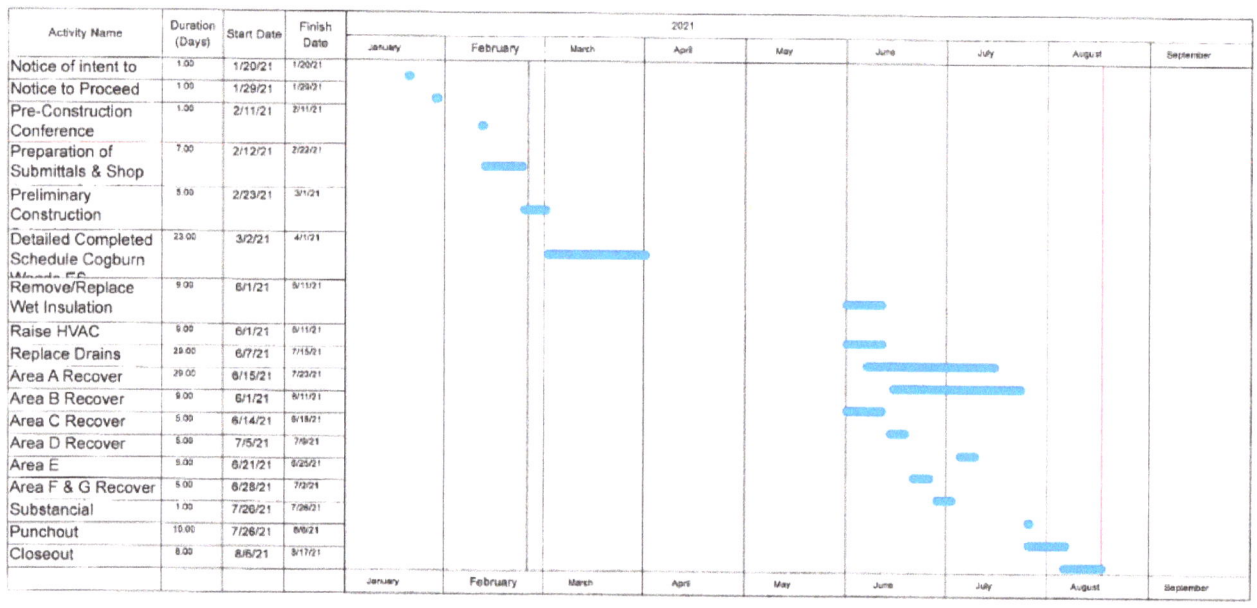

| Activity Name | Duration (Days) | Start Date | Finish Date |
|---|---|---|---|
| Notice of intent to | 1.00 | 1/20/21 | 1/20/21 |
| Notice to Proceed | 1.00 | 1/29/21 | 1/29/21 |
| Pre-Construction Conference | 1.00 | 2/11/21 | 2/11/21 |
| Preparation of Submittals & Shop | 7.00 | 2/12/21 | 2/22/21 |
| Preliminary Construction | 5.00 | 2/23/21 | 3/1/21 |
| Detailed Completed Schedule Cogburn Woods FC | 23.00 | 3/2/21 | 4/1/21 |
| Remove/Replace Wet Insulation | 9.00 | 6/1/21 | 6/11/21 |
| Raise HVAC | 9.00 | 6/1/21 | 6/11/21 |
| Replace Drains | 29.00 | 6/7/21 | 7/15/21 |
| Area A Recover | 29.00 | 6/15/21 | 7/23/21 |
| Area B Recover | 9.00 | 6/1/21 | 6/11/21 |
| Area C Recover | 5.00 | 6/14/21 | 6/18/21 |
| Area D Recover | 5.00 | 7/5/21 | 7/9/21 |
| Area E | 5.00 | 6/21/21 | 6/25/21 |
| Area F & G Recover | 5.00 | 6/28/21 | 7/2/21 |
| Substancial | 1.00 | 7/26/21 | 7/26/21 |
| Punchout | 10.00 | 7/26/21 | 8/6/21 |
| Closeout | 8.00 | 8/6/21 | 8/17/21 |

Figure 4.5.2–Roofing Schedule

The following is another example of a project, this time using MS Project software.

This project involves installing 100,000 sf of TPO membrane with 2.2" of polyisocyanurate roof insulation and 3' flashing for 800 linear feet of parapet wall.

The roof is in two different sections, one 75.000 sf with 600' of wall flashing; the other is 25,000 sf with 200' of wall flashing. The project is to be performed in two phases, and the existing roof needs to be torn off. One crew consisting of a foreman, two mechanics and two laborers/helpers will be used on the project.

Phase I includes:

- Mobilization (2 days)

- Tear-off existing roof (10 days)

- Surface preparation (3 days)

- Installation of barrier, insulation and membrane (62 days)

- Installation of flashing (5 days)

- Handling punch list items (3 days)

Phase II includes:

- Tear off the existing roof (4 days)

- Surface preparation (1 day)

- Installation of barrier, insulation and membrane (20 days)

- Installation of flashing (2 days)

- Handling punch list items (3 days)

- Demobilization (2 days)

The days for barrier, insulation and membrane installation for Phase I have been calculated using the same method used in the previous example:

$$\frac{75,000 \text{ SF}}{30 \text{ SF}} = 2,500 \text{ MH}$$

$$\frac{2,500 \text{ MH}}{40 \text{ CH}} = 62.5 \text{ Days}$$

A similar method of calculation would be used for Phase II. It would be calcuted as:

$$\frac{25{,}000 \text{ SF}}{30 \text{ SF}} = 833.34 \text{ MH}$$

$$\frac{833.34 \text{ MH}}{40 \text{ CH}} = 20.8 \text{ Days}$$

Figure 4.5.3 is an example of all the necessary information entered into MS Project. Notice the column named "Predecessors." This column is used to link various activities. For example, the tear-off activity has been assigned Activity 2 as a predecessor. This means that tear-off cannot take place unless mobilization has been completed. Similarly, for surface preparation, the predecessor is 3FS – 2 days, meaning the activity will start 2 days before the tear-off finish date. On the right-hand side, a Gantt Chart is created using the information on the table on the left-hand side.

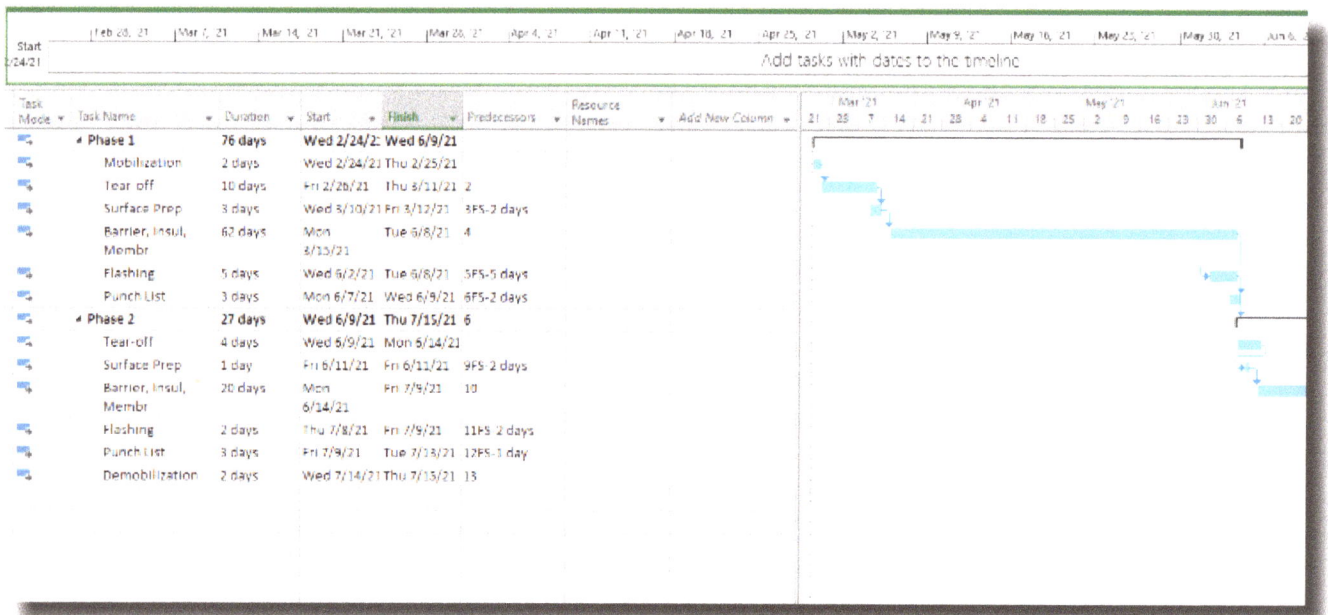

Figure 4.5.3– Roofing Schedule

Insurance is a risk management tool that is a contractual agreement between the policyholder and the company providing insurance – which means it is valid only during the contract period. The policyholder, or insured, receives a policy detailing the terms and conditions of the insurance and pays the insurance company a regularly scheduled payment, called a premium, for the risk coverage outlined. In return, the insurance company provides coverage and will pay claims that occur in accordance with the policy. There are many types of insurance, such as health, vehicle, house, life, and business insurance. Insurance takes on different roles according to the stakeholders involved. Some stakeholders in construction and roofing include Owner/ Developer, Architects/Engineers, Lender/Bank, General Contractor, Roofing Contractor, and Roofing Manufacturer. Some common types of insurance in the industry include:

## Builder's risk

Coverage for damages or losses to a project due to fires, theft, or vandalism during construction. Builder's risk is provided based on factors such as the complexity of construction and how the security of high-value materials is ensured on the project.

## Worker's compensation

Coverage for employees injured or ill due to work. Worker's compensation is provided based on factors such as the type of work, the project's complexity, the work's physical demands, and the safety program in place. Additional factors that are likely to influence the premium include training hours, the number of employees, the nature of work (high-risk vs. low-risk), the safety record of the company, and previous incidents.

Coverage for errors, omissions, or mistakes made by architects, engineers, or other professionals on the project. Professional liability is provided based on the design's complexity, the work's value, and the team's credentials.

Coverage for injuries or property damage during construction. General liability becomes a part of most policies.

Coverage for damage, theft, or liability on the project. Liability coverage is provided based on on-site hazards, engineering controls, and administrative controls. Theft coverage is based on on-site high-value items, security systems/procedures, location, and emergency response time. Damage coverage is provided based on adherence to building codes and vulnerability to risk (vandalism/break-in, high wind events, hailstorms, and fires).

Fortified is another building standard that includes voluntary, beyond-code requirements. It is developed by the Institute for Business and Home Safety (IBHS). It provides a set of standards that a building must meet to ensure resiliency against high winds, whether inland or on the hurricane coastline. A fortified building designation showcases the safety and resiliency of a building and can result in lower insurance premiums.

As shown in Figure 5.1.1, the roof is the number one line of defense against inclement weather conditions. Five of the ten elements IBHS has identified to reduce hurricane risks in the diagram pertain to the roof: roof condition, perimeter flashing, roof-mounted equipment, skylights, and lightning protection systems.

Figure 5.1.1—IBHS Guide to Reduce Hurricane Risks

## 5.2 Risk Management

Every construction project comes with its own set of risks. Any deviation from the planned project outcomes regarding people, property, cost, schedule, or quality is a risk for the project. Risks, most often, cannot be eliminated. However, they can be minimized and managed.

There are risks inherent to every project. High-risk projects provide a higher rate of return. However, it is essential to understand the risks the company is equipped to manage. It is wise to reject projects of much higher risk than the company can bear.

There are two phases to risk: potential and actual. Potential risks have not happened yet but could potentially occur during construction. Actual risks are ones that are in progress or are happening during construction. The best phase to be is in the potential phase, where risks can be identified prior to construction, and a risk management plan can be devised. In this way, the impact of the risk, should it actually occur, can be reduced. If a risk not identified during the preconstruction phase occurs, an immediate action plan must be created to minimize its impact on the overall project.

Risk identification is the first and most crucial step in the risk management process. One should be objective, realistic, and prudent when identifying risks. Once risks are identified, they need to be measured in terms of the following:

- Probability – Likelihood that the event will occur during construction
- Frequency – How often the event is expected to occur
- Severity – Amount of possible loss due to the event
- Timing – When the event is likely to occur

Risks must be identified early and addressed. They must never be ignored. All possible impacts of risk must also be identified. Following this, the ability to manage the risk must be evaluated. Contractual conditions and obligations must be evaluated before developing a risk management plan.

Two major risk categories on a construction project are global and project-specific risks. Global risks generally affect every project. They are difficult to assess, especially for long-duration projects, but could still be assessed before and during the contract. Some examples include economic activity, inflation/deflation, acts of god, war, civil disruptions, material shortages,and labor strikes. Project-specific risks are inherent risks on a project.

It is primarily the Project Manager's responsibility to identify these risks. The Project Manager assumes the role of a "Risk Manager" on every project. These risks should be assessed and incorporated into the project planning and execution phases. The company can either minimize them, or the Project Manager can assign the risk to another appropriate party equipped to handle it.

There are four ways of looking at project-specific risks:

1. Owner type
2. Design/engineering/consultants
3. Subcontractor
4. Job site conditions.

The overt involvement of an owner, quality of the Architect and Engineering (A&E) team, response time to submittals and Requests for Information (RFIs), subcontractor activities on the project, and safety of the job site are typical sources of project-specific risks.

Experienced Project Managers or vendors have a "vision." They can envision the project from the beginning to the end. In doing so, they anticipate any possible risks they might encounter in their scope of work and are prepared to manage them.

# CHAPTER

# 06

Site Logistics

Site logistics refers to planning and managing activities and resources on a construction project, including material and equipment procurement, inventory management, material handling and the movement of people. It can be defined as "setting the stage for project success," because the site logistics plan is vital to ensure the project is executed efficiently and meets its deadlines with minimal disruptions.

Key logistics considerations on a construction project include safety and health, production and security. It is important to ensure the safety of workers, visitors and the general public on every project. The site logistics plan must consider potential hazards involved in transportation, material handling, equipment operation and other site activities.

Another important consideration is the health of the workers. Access to the site, walk paths, noise management, air and water quality, ergonomic issues and toilet facilities all contribute to the overall health and well-being of workers on the job site. Production, which includes job sequencing, production rates and best methods, is another consideration to keep in mind. It is also essential to ensure the security of the job site through site surveillance, monitoring and site access controls.

The site logistics plan depends on the project type – and, in particular, whether the building is residential or commercial.

Access points are critical for residential projects; they differ depending on the following:

- The type of roof

- The type of building – single-family or multi-family

- The height of the structure

- Ground conditions

- Other trades on the project – painters, electricians, solar installers, etc.

Commercial projects have the same considerations as residential projects, but to varying degrees and for different reasons. Figure 6.1.1 is an example of roofing a commercial building.

Notice there are two cranes on the project – the roofing contractor's and the general contractor's. It is important to ensure that the operation of different pieces of equipment on site do not clash or cause safety hazards for anyone on site. Notice that access to the roof is by a man lift. The building height becomes a more important consideration in commercial projects. Working on a tall tower in a busy neighborhood is always a challenge.

Coordination and communication also become important considerations as more trades might be simultaneously involved in the project. Complex commercial projects need to have a higher level of planning and coordination. Figure 6.1.2 is an example of roofing a typical low-slope commercial building. Notice the work sequencing from one end of the roof to the other and safety features such as cones and warning lines.

Figure 6.1.2 – Low-slope Commercial Building

## 6.2 In-bound and out-bound operations

In-bound logistics refers to materials, equipment, personnel and vehicles entering the construction site. All need to be brought to the appropriate location–safely and on time– for the work to commence. Some key considerations include the following:

- Equipment selection. Selecting the correct type of equipment needed to get materials and workers onto the roof.

- Scheduling. Resources need to be available on time so the crew can begin working.

- Material loading.

- Ground safety.

Figure 6.2.1 shows a crane that is used to hoist materials onto the roof. The area below the crane has been secured with caution tape to restrict access to people on the ground and to ensure ground safety. A spreader bar is used to evenly spread the load being lifted and to prevent buckling of the metal panel. Figure 6.2.2 is an example of how ground safety is ensured.

Important considerations for out-bound logistics include material handling, removal of debris and refuse, sanitation, recycling requirements (separating, bundling and storing materials) and sustainability concerns, such as handling chemicals and volatile organic compounds (VOCs). Figure 6.2.3 shows how a trash chute is used to discard roof waste into dumpsters. The ground area surrounding the chute and the dumpster is fenced to ensure that passersby are safe from falling debris.

Figure 6.2.4 is an example of how a vacuum truck is efficiently used to remove gravel from the roof's surface. A typical reroofing project produces a large amount of waste as the old roof system is torn off. An example is shown in Figure 6.2.5.

Figure 6.2.4–Vacuum Truck for Gravel Removal

Figure 6.2.5–Roofing Waste

## 6.3 Job Sequencing

The first consideration in out-bound sequencing is how people and materials are removed from the roof. Different access systems include scaffolding, ladders, stair towers, lifts, buck hoists and other similar equipment.

Figure 6.3.1 shows an example of scaffolding and a buck hoist (in temporary elevation) used for vertical movement. Notice that a crane is also used to hoist materials onto the roof and to bring discarded materials down to the ground.

Figure 6.3.1–Access Systems

Job sequencing sets the stage for the most accessible work path. Materials should be placed to allow for easy access and installation by the crew. Job sequencing also involves coordinating with other trades working on the roof; the idea is to minimize handling and create efficiencies with planning, strategic material placement and planning for removal. Loading and staging of materials should allow for efficient completion of work and meeting daily installation goals. It is important to stage the proper amount of materials that will be installed within a set period of time.

Site logistics planning is often influenced by weather conditions, such as hurricanes, snow, ice, strong winds, tornados and extreme heat. Protecting and safeguarding materials becomes more important if harsh weather conditions have been predicted. Figure 6.3.2 is an example of a project in Tampa, Fla., where a sudden hurricane warning was issued. Since there was not enough time to get materials off the roof, it was decided to double-bag all stacks of materials. Notice that tarps have been wrapped around some equipment and bolted to the concrete deck to ensure they stay in place. Since there was not enough time to get materials off the roof, it was decided to double-bag all stacks of materials. Notice that tarps have been wrapped around some equipment and bolted to the concrete deck to ensure they stay in place.

The lack of a good logistics and access plan can lead to significant safety hazards and can cause delays, inefficiencies, increased costs and reduced productivity.

Rooftop safety demands that the contractor consider a whole host of issues, including:

- Personal protective equipment (PPE)

- Fall protection

- Protection from dehydration and heat illness

- Protocols from OSHA and the Centers for Disease Control and Prevention (CDC)

- Fire prevention

- Equipment operation, which may involve using generators

It is also crucial to consider access and egress, including how the placement of access systems might affect activities on the ground and surrounding facilities. There might also be pedestrian and ground traffic immediately below the work area; hence, debris falling from the roof must be contained.

Because roofing is a hazardous trade, largely due to working at heights and slopes, safety must be incorporated at every stage of the logistics plan. Failure to plan properly can lead to catastrophic injuries and even fatalities. An inefficient and unplanned project will most likely reduce profit margins. An accident or injury will result in increased insurance premiums, damage to the company's reputation and loss of morale and productivity with the crew. Most likely, the customer will lose confidence and trust in the company's work. Safety planning is crucial for every roofing project.

Field Crew Management

In the early days of the roofing industry, crew members were hired and paid based on the quantity of work they completed; this was known as "piece work." The pay was determined by the number of "squares" – 100 square feet equals a square – they installed. This has led to the rise of crews that are independent of the primary roofing company, which, among other things, takes scheduling flexibility away from the contractor. To a certain extent, it also took control of workmanship and safety away from the contractor, even though the contractor was ultimately responsible for all that happened on the job. Too often, this led to an unsafe work environment and substandard work product, damaging the profession's credibility.

The model of quantity-based pay still exists, although the industry is learning that in-house (employees of the contractor) crews paid by the hour have many advantages. They use company trucks, interact daily with company personnel and with each other, helping to build trust. They are also trained by the company, enabling new hires to learn from experienced journeymen. This more interactive model leads to more consistent roof installation, better communication, better training, easier supervision and more control over the work product. The availability of trained personnel increases efficiency for the company by providing the flexibility to manage the crew size according to specific job needs. Planning, training and communication are essential for efficient crew management.

## 7.2 Crew Planning and Monitoring

The first step in planning for a roofing project is estimating. The estimator must thoroughly understand the project's scope and plan for all of its phases. The estimator must consider such things as the size of the crew required, equipment needed, site logistics, transportation and travel.

The estimator must also know the height and slope of the roof, the geographic location of the building, any unusual site conditions, the type of project (new construction or reroofing), surrounding buildings, traffic and interaction with the public. Site visits and conversations with the owner and suppliers are crucial to arrive at the best practices for the project.

The schedule and budget are important to project planning. For new construction, the general contractor typically provides an expected time frame, including sequencing for the project. However, there may be delays from prior trade work or the weather that could impact the schedule. It is common to see projects with a strict deadline that need the roofing work to be done in less time than the roofing contractor initially planned. In those cases, the roofing contractor might have to increase the number of crew members or the length of the shifts they work. This may also result in additional costs, so the lines of communication between the general contractor and the roofing contractor must always remain open.

Once the initial planning phase of the project is completed, careful consideration must be given to the field crew, including its size and composition. The size and composition of the crew often depend on the project's complexity. The skill set of the crew must align with the requirements of the roofing work.

Consideration must also be given to whether the crew is capable of operating the equipment that will be used, and they must also be able to handle any unique safety challenges associated with the project. Often, training must be factored into the planning process, especially when new equipment is used or when the crew is unfamiliar with the roofing materials to be installed.

Crew members must be briefed about the overall scope of the project and how the work must be approached. Preconstruction meetings should be held not only to explain the project plan but also to get the crew on board. Any concerns or discomfort about the plan brought up by crew members must be brought up during this meeting. Managing a field crew depends on communicating the plan, getting buy-in and maintaining open communication. One crew member who is not aligned with or doesn't understand the project plan can cause disruptions to the entire chain of operations and reduce the crew's effectiveness.

The first day of work is the most important day. It sets the tone for productivity and organization for the rest of the project. A supervisor must be present on-site to ensure that all necessary resources are in place and that the crew is aware of what their activities will be. The first day is also used to ensure that all crew members understand the plan. Organizing the crew members – by briefing them about their roles and their importance – can help with their approach to the project and their productivity.

After the project begins, the next step is monitoring the progress of the job against the plan. This includes analyzing the job costs, production levels and ongoing challenges that occur in the course of the job. Spreadsheets are often designed and used to tracks actual work and budgeted costs for labor and materials for each project. A detailed report that includes budgeted, actual and projected expenses can be produced. This can and should be monitored in real-time. These types of reports make it easier to take corrective measures when a project is running over budget. Additionally, photographs of the work in progress are essential for reference post occupancy if there are issues with maintenance, replacement, warranty, or even litigation.

Photographs help to track milestones, identify issues and delays, assess the quality of the work and communicate with the owner and other stakeholders. This documentation also proves to be valuable should a question arise as to where the progress was at any given point in time.

Unforeseen circumstances and inclement weather will influence the project and its schedule. It is crucial to account for these conditions and allow for "floating days" in the project's schedule. For instance, if severe weather is expected during a roofing project, the crew may need to be reassigned to a project with an immediate emergency need. Also, projects do not always go according to plan, so the flexibility to change a crew's size or makeup or work additional hours or days is essential to meet unforeseen circumstances.

It is important to acknowledge that crew members are people with other obligations and responsibilities in their lives. They may fall sick, have family emergencies or have other reasons that can pull them away from their work.

It is important to recognize, understand and support the well-being of the crew, and supervisors will have to be able to schedule changes as needed and communicate them to the rest of the crew and the owner.

With good planning, communication and training, the field crew will be positioned for success. Teamwork is always critical, as the crew is a team that must work together to meet the project's goals. Conflicts and issues can arise, of course, because crew members are people with different personalities. When conflicts arise, it is important for all crew members to be listened to and for the supervisor to attempt to find common ground. A good supervisor will nurture a positive working relationship among crew members and customers to achieve successful completion of the project.

# CHAPTER
# 08

Roofing Equipment

The role of roofing equipment in the past, present and future remains the same – to reduce labor costs, increase worker safety and deliver a better quality of work. Roofing is a labor-intensive activity and requires constant exposure to the weather. The use of equipment is therefore essential to ensure that the best results can be achieved in a limited period of time.

A roofing contractor primarily focuses on delivering quality projects on time and on budget. Using the proper equipment is certainly a good way to improve productivity. Equipment used in roofing can vary from small hand tools to large cranes; as the roofing industry has changed, new equipment has been developed to meet the needs created by new materials and new code and regulatory requirements.

Built-up roofing (BUR) was introduced in the 19th century and quickly became widely adopted for low-slope roof systems. BUR gets its name from the way it is applied – layers of asphalt are applied between layers of roofing felt (or tar paper), building up the system on the roof. Usually, BURs are finished with a top layer of gravel or a reflective coating. It is a very labor-intensive installation process. The methods – and equipment – used for installing BUR changed very little until the 1930s when equipment manufacturers introduced new equipment to heat bitumen.

Following World War II, the demand for low-slope roofing boomed as the country built schools, hospitals, manufacturing plants and the like. This pressed the industry into finding new ways to mechanize low-slope roofs and boost productivity.

Today, there are many roof systems on the market to meet clients' requirements. The majority of the work done in the industry is retrofitting–replacing a roof or "reroofing."

Roofing maintenance, as well, is becoming increasingly important to sustaining the life of low-slope roofs. Many roofing contractors specialize in one type of roofing, e.g., metal, single-ply or BUR. Others perform work on many different types of systems. Specialized equipment for different roof systems has been designed to make work safer, faster and of higher quality.

## 8.2 Safety Equipment

Roofing is considered one of the most dangerous construction trades, even though it is much safer today than it used to be. This is due to a number of factors, including the advent of government regulation (OSHA) and the role of the insurance industry in insisting on better work practices and better safety equipment.

Roofing still requires working on elevated surfaces, working under all sorts of weather conditions and with uncertainty about the condition of the building getting a new roof. But contractors today focus much more on safety for a lot of reasons, including safe operations are usually cleaner and of higher quality and safe workers are more productive and feel valued. Hence, contractors today are more willing to invest in safety equipment and safety training than ever before. It is also true that some owners will only allow contractors with good safety records to bid on their work.

In the absence of some kind of perimeter protection (e.g., scaffolds or safety nets), OSHA requires workers on steep-slope roofs more than 6 feet off the ground to have fall arrest systems, usually composed of a rope and lanyard system.

Most body harnesses are designed to withstand 5,000 pounds of pressure. More connection points on the harness indicate better quality and design. The placement of connection points is crucial to providing good support for workers and reducing the likelihood of injury should they fall. An example is shown in Figure 8.2.1. Harnesses are an important part of a fall arrest system and require that users be properly trained.

Figure 8.2.1 – Body Harness

A retractable lifeline, which contains a steel cable that can be extended and retracted as needed helps reduce the stress on the harness and, hence, on the worker if a fall occurs. However, lifelines do require regular inspection and maintenance to be sure the braking mechanism performs as intended. Dirt and debris can impact their performance, so it is important that they be stored correctly. Retractable lifeline with a swivel attachment are available. Additionally, permanent anchors for the lifeline system building onto the roof structure help with safety during maintenance, inspections or reroofing work, especially on steep roofs. Some codes have this as a requirement.

Slide guards can help prevent falls on steep-slope roofs and can also help to keep tools and materials in place. OSHA used to allow their use as the sole means of fall protection but now requires full fall protection systems of harnesses and lanyards.

"Warning lines" are allowed along the open edge of most low-sloped roofs. As shown in Figure 9.2.7, warning lines let workers know they are nearing the roof's edge; the must usually be placed no closer than 6 feet from the edge, and they must be able to withstand weight to keep the supporting stanchions from tipping. When workers are outside of the warning lines, a safety monitor must be present to ensure the workers' safety.

Figure 8.2.2 – Warning Line System

Portable achorage systems for steep-slope roofs are devices that can anchor multiple workers; however, it limits work to one side of the roof at a time. Portable anchorage systems can also be used on low-sloped roofs, often to protect workers who are beyond the warning line system and the edge of the roof. In such cases, a portable anchorage cart can be used. Some carts are designed for fall restraint only. They only allow the worker to reach over to the edge of the roof. Other carts that provide fall arrest give workers more freedom of movement.

Guardrail systems for are restraint systems that allow work to proceed without using restraint systems such as ropes or harnesses. On low-sloped roofs, they are used instead of warning line systems. Trained, competent people need to be involved in installing these systems; however, workers on the roof do not require any special training. These systems can be temporary (Figure 8.2.3) and there are proprietary systems available that allow for easy installation and reuse from project to project.

Permanent guardrail systems are less common and generally only installed by roofing contractors if they are part of the specifications or required by code.

Common equipment for reroofing operations includes tear-off spades, shingle removers, spud bars, rotary planers and roof cutters, among others.

A single-blade roof cutter is another piece of equipment that makes the removal of built-up roofs much more convenient. It is widely used on commercial reroofing projects. The hand crank sets the depth of the blade and can be easily adjusted.

Loose gravel on built-up roofs must be removed prior to the removal of the rest of the roof. Power sweepers are used for this purpose. For larger projects, roof vacuuming is done. Because roof vacuums are costly to use, maintain and store, companies that provide vacuuming services to roofing contractors are commonly used.

Powered roof removers work best for built-up roofs that are adhered to concrete decks. They can be used on other types of roofs as well, but they may be too heavy to use on lightweight metal decks. The oscillating blades in this machine also pivot, which helps on roofs with rough or uneven surfaces.

As a general rule, most roofing contractors – unlike general contractors and some other specialty contractors – do not typically rent equipment. Roofing contractors prefer to own their own equipment, but the introduction of large and expensive equipment, such as cranes and roof vacuums, has resulted in more equipment being rented.

Trash chutes are commonly used to contain and lower roofing debris on reroofing projects. Chutes are composed of high-density plastic and shipped flat with metal strips on the ends. These metal strips are bolted together on site and unbolted for storage. The chains on the chute connect to the chute above it, which eventually connects to a hopper arrangement at the top.

The pockets on the hopper's side are used to firmly attach the hopper to the roof. Figure 8.3.1 shows an example of a trash chute in a mid-rise building. Erecting the trash chute with scaffolding is common to provide additional support.

Figure 8.3.1: Trash Chute

A trash "rocket" is another new type of equipment for low-slope re-roofing. It is must faster to install than traditional chutes and is particularly helpful in cases where it is impossible to get close to the building. An example is shown in Figure 8.3.2. The trash rocket comes with a truck or trailer-mounted options.

Small engines are used on a lot of roofing equipment, and is important to remember that they require regular maintenance. If an engine runs out of oil, for example, it can result in significant delays and might also damage the engine itself.

Protecting landscaping around a home or building during a reroofing project is part of the roofing contractor's responsibilities. A new system, called the "Catch-All," is designed to protect plants and shrubs while collecting nails, metal fragments and other debris. An example of the system is shown in Figure 8.3.3.

Figure 8.3.3– Catch-All System

8.4 Material Handling Equipment

Two-wheel wheelbarrows are commonly used at virtually every job site. Wheelbarrows made for use on roofs typically have a heavy-duty tubular frame and a nose that is friendly to roof membranes.

A hoisting wheel is often the best way to get a few buckets of cement or coating to the roof. Hoisting wheels are used extensively for roof maintenance work.

On the roof, four-wheeled utility carts are also commonly used. These carts can be pulled manually or by a powered tractor. They can also be used with several different attachments.

Manual forklift dollies are commonly used for handling roof insulation. Roof insulation is bulky but lightweight and requires careful handling. There are also power forklifts designed for use on the rooftop. It comes with pneumatic tires and a wide stance.

Manual swing-boom hoists are not commonly used for lifting material to the roof but are ideal for long-term projects requiring lots of small loads.

Track and trolley hoists are powered by small gasoline engines and can be used with accessories such as gravel hoppers and slings. Roof-mounted hoists are slower than belted conveyors but can be used if there are not too many workers or equipment on the roof. A hydraulic swing-boom hoist is an upgrade to the manual swing-boom hoist due to its smoother operation and larger capacity.

A telescopic elevator is towable and telescopic, as the name suggests. It is not commonly used in the U.S. However, some contractors use these elevators to hoist material and lower debris. They are used more frequently in Europe because they can be set up in confined spaces. Truck-mounted versions are also available but are mostly used when a project requires rapid deployment among a number of buildings.

Rough-terrain forklifts, as shown in Figure 9.4.14, are called "all-around" lifts because they are used to lift and lower any type of material and are commonly used by all trades on a new construction project. Operators of forklifts must be trained; however, certification is not required in all jurisdictions. A telescopic lift with a man-cage, as shown in Figure 9.4.15, is commonly used for roof edge work or metal roofing.

Figure 8.4.4 – Rough Terrain Forklift

Figure 8.4.2 – Telescopic Lift with Man Cage

9.5 Built-up Roofing Equipment

Built-up roofs consist of layers of alternating layers of roofing felt and hot asphalt. The asphalt can be applied by hand-mopping or by using a machine called a felt layer. For hand-mopped applications, a mop and a cart are standard equipment, along with another piece of equipment called a "hot lugger," to transport the asphalt to the mop cart.

Some modified bitumen roof systems are also applied using hot asphalt, as shown in Figure 9.5.1. Modified bitumen rolls are essentially pre-manufactured built-up roofing rolls, with the first layers produced in a factory and the top layer applied on the roof.

Cotton mops, used to apply the hot asphalt, have been used for a long time but are being replaced by fiberglass mops, which have a much longer useful life.

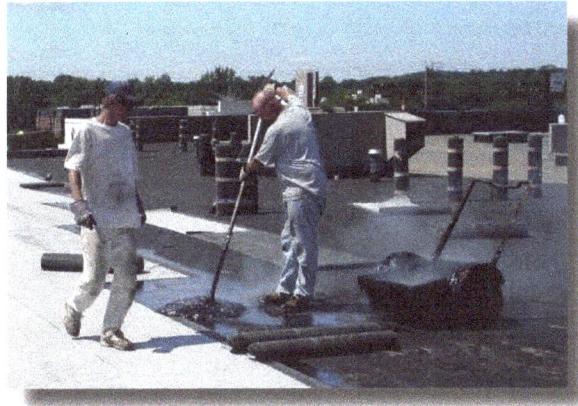

Heating the asphalt for both built-up and modified bitumen roofs is usually done using equipment on the ground. A smaller "patch kettle" is used to heat asphalt on the job site; a smaller version was developed in the 1970s with the advent of modified bitumen roofs.

A pumper kettle has the kettle mounted onto a trailer with a pump that moves the

hot asphalt to the roof. An example is shown in Figure 8.5.2. The trailer has an 8 or

9-horsepower engine. Earlier versions were made with the pump detached from the

kettle.

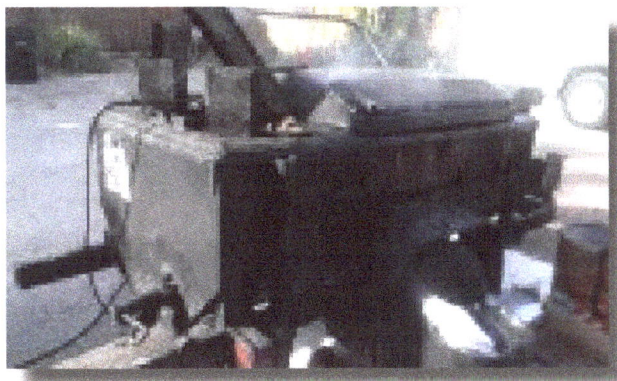

A day tanker delivers hot asphalt to the roof with a pumping system similar to a pumping kettle. It is designed for both transport and storage of the asphalt. The truck has large insulated holding tanks that hold 7 to 15 tons of asphalt. A propane heating system, controlled by a thermostat, is used and eliminates the need for a kettle operator, as well as the costs of handling packaged asphalt. If asphalt is unavailable locally, the contractor often elects to set up their own yard storage systems. An example of a truck-mounted day tanker is shown in Figure 8.5.3.

Figure 8.5.3 – Liquid Asphalt Tanker

As shown in Figure 8.5.4, a hot lugger is used to transport asphalt on the roof.

Figure 8.5.4 – Hot Lugger

A felt layer is commonly used on built-up roof projects today. Roofing felts are loaded onto the machine, which the operator pushes over the roof. As the felt layer moves, the roll is unwound and applied in a layer of hot asphalt. An example is shown in Figure 8.5.5.

Figure 8.5.5 – A Locomotive Felt Layer

A manual gravel spreader is used to spread gravel across the surface of the roof. The machine spreads gravel or slag evenly and embeds it into the hot asphalt. A powered gravel spreader does not operate more quickly than a manually operated one but is much easier on the operator and can be used for longer stretches.

A hot dispenser (Figure 8.5.6) is mainly used to apply hot asphalt in between layers of roofing felt. It is less expensive than a felt layer but requires the felts to be set by hand. A mini-mopper is a low-cost version of a hot dispenser. The tiny holes on the front of the mopper dispense asphalt while the attached chains drag over the asphalt.

Figure 8.5.6—Hot Dispenser

8.6 Modified Bitumen Equipment

A machine called a Dragon Wagon, as shown in Figure 8.6.1, is commonly used for applying rolls of modified bitumen. It is used for torch-applying modified bitumen felts onto the roof substrate or base layer. Notice the torch heads attached to the equipment. One handle controls the torches (flame setting and orientation), and the other controls the rolls of modified bitumen.

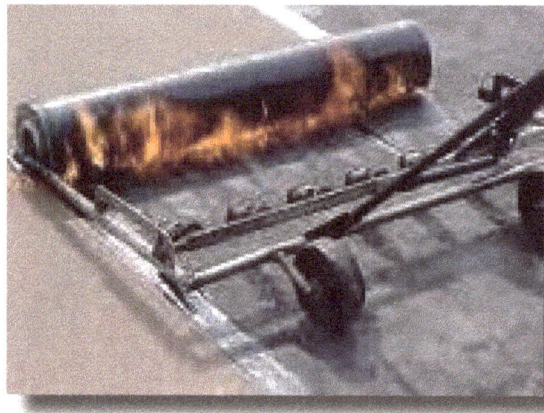

Figure 8.6.1—Dragon Wagon

Torching is also done manually, using a hand torch. Hand torches are propane-fueled with flame-producing nozzles. A typical hand torch outfit consists of two sizes of torches with trigger handles, strike, hose, and regulator. Larger torches are used for torching full rolls and smaller torches are used for detailed work.

A Mini McCadden modified bitumen welding machine uses heated air instead of an open flame. This feature keeps the roll of track automatically while it reduces the risk of fire.

The use of torching equipment with open flames obviously presents some fire hazards, especially when torches have been used near flammable materials, such as wood blocking or near open vents.

There have been many cases when fires have started up to an hour after the roofing work was completed for the day. It is a good idea to take temperature readings of the installed membrane and to have one or more observers on the roof for an hour after the day's work has been completed.

The National Roofing Contractors Association offers a program called CERTA – Certified Roof Torch Applicator – to train torch operators about the best practices to prevent fires on the roof. Many specifications and some insurance companies require CERTA-certified torch applicators.

## 8.7 Spray Foam and Coating Equipment

Spray foam roofing involves applying a two-part liquid foam that, once sprayed, expands and hardens to form a seamless, waterproof barrier. The application of sprayed foam – and roof coatings as well – involves a variety of equipment, including pumps, heaters, pressure regulators and heated/insulated hoses. The application of sprayed foam requires expertise, and a spray foam operator must also use appropriate PPE. A spray foam rig can be used for various roof coating applications, such as spraying metal roofs. Another piece of equipment is the powered roller outfits. This is used by roofing contractors who prefer roller application to spraying.

Attaching the seams of single-ply roof systems correctly is key to their success. One method for seaming single-ply roof membranes is heat welding, and a device commonly used for that purpose is a hand-held heat gun (figure 8.8.1). The hand-held heat gun is primarily used for welding thermoplastic membranes such as TPO and PVC. Small tools are needed for welding applications; these include a brush (for cleaning), silicone roller, fine detail roller, test pick, shears and a pair of nozzles.

Figure 8.8.1– Robotic Welders, Hand Welders, and Accessories

A robotic welder is not completely robotic and does require an operator. It is important to note that both hand guns and robotic welders need to be cooled down (for at least 10 minutes) before they are shut off for the day.

Another way to attach seams on single-ply roof systems is through the use of an induction welder with magnetic heat sinks (Figure 8.8.2). The Rhino Bond welder, manufactured by OMG Roofing Products, uses induction weld technology to adhere the roof membrane to the substrate. Once the membrane is laid out on the roof, this device at placed at specific intervals to send a high-frequency electric current through the heat sinks. This causes heat to build up between the membrane and the substrate, leading to the bonding of the membrane with the substrate.

A stand-up screw gun, shown in Figure 8.8.3, has a stack of 100 plates on it. The plates, with screws, are loaded into the tube in the front of the screw gun. The pedal and rollers allow the screw gun operator to move along the roof, installing screws. This is a much faster method than using a typical screw gun.

Some single-ply membranes are manufactured with a layer of fleece attached to the underside of the membrane for additional reinforcement and puncture resistance. These membranes are attached to the substrate using adhesives. Fleeceback rollers are typically used to apply even pressure over the membrane as it sticks to the adhesive.

Generators are typically used on virtually all single-ply roofing projects in order to power robotic welders, heat guns and other equipment on site. There are a number of different types of generators with various electric outputs.

## 8.9 Metal Roofing Equipment

A number of tools are used to install metal roof systems. Because metal is not very malleable, it requires different tools for cuts and seam work. As shown in Figure 8.9.1, the first-stage seamer is used on a standing seam metal roof to bend or fold edges. A second-stage seamer, shown in Figure 8.9.2, is used with a first-stage seamer to finish seams, creating a watertight seal.

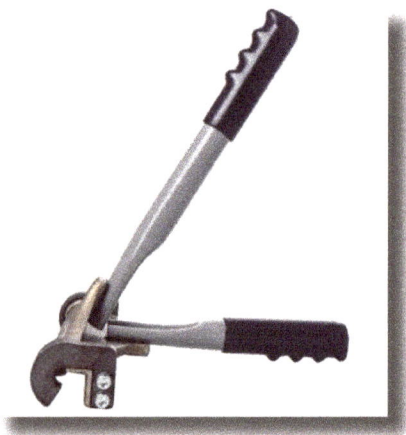

Figure 8.9.1 – First Stage Seamer

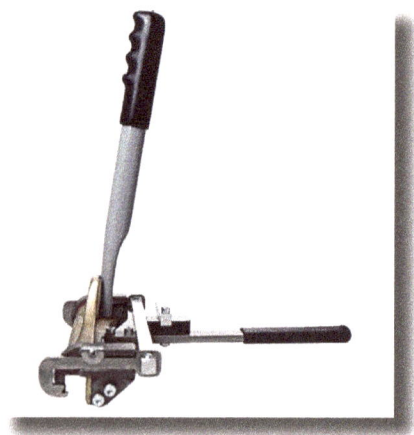

Figure 8.9.2 – Second Stage Seamer

A cordless power seamer, shown in Figure 9.9.3, is a mechanized version that functions similarly to the first- and second-stage seamers. Cordless power seamers are portable and convenient to use and run on battery power. A power seamer, like the one shown in Figure 9.9.4, has multiple rollers and works faster than manual seamers.

A portable roll former, shown in Figure 8.9.4, makes metal panels for roofing on site. It consists of rollers that shape metal coils into panels. They address many production issues with metal roof systems, most importantly the logistics of getting metal panels to the job site. These machines come in a variety of sizes and styles and can produce panels at virtually any length.

A platform hoist is a gas or electric-powered piece of equipment used to take shingles to the roof. It is lightweight, portable, inexpensive and easy to erect.

A truck-mounted shingle and tile conveyor is used to get materials to the roof. These conveyors were introduced in the 1970s; the latest innovation is a telescopic conveyor that reaches up to 53 feet.

A shingle cutter is commonly used; it is especially useful for working on domes.

Roof brackets are attached to the roof to provide roofers with extra support and stability, especially when working on very steep roofs. Pivit ladders provide a flat surface on a sloped roof. They are attached to ladders and provide a base for roofers to stand on or to place tools.

Air-powered nail guns use compressed air to nail or staple shingles to the roof deck. They are widely used. Air-powered nail guns also require a compressor on the job site. Another option is the cordless electric nail gun. These are heavier and more expensive than air-powered nail guns and, therefore, not commonly used; however, they do not require the compressor or have the hazard of air lines being tangled in the work area.

Applying a slate roof requires that the slate be cut and hung. There are different types of slate cutters available today that include hand slate cutters (8.11.1), cordless electric slate cutters (8.11.2), and workbench mounted slate cutters. Pieces of slat need to have holes in them so they can be hooked on nails. These holes are made using a tool called a slater's hammer.

A slater's skate, shown in Figure 8.11.3, is another tool that can be used to puncture a hole in the roof deck to attach nails to the slate. A slate ripper is a tool used to remove nails to replace existing slate or asphalt shingles. The end of the slate ripper is slipped under the shingle to remove the nails.

Roof cutters are used by contractors and consultants to conduct roof inspections. Core cuts–small samples taken by cutting into the roof–can provide important information about the roof system and can help to identify problems.

A Rolatape measuring wheel is a tool used for measuring. A wheel is attached to a pole or handle that has a counter displaying the distance the wheel covers.

A foam patty, also known as a foam spacer, is placed on the underside of a concrete roof tile to prevent the tile from shattering during high winds and to provide ventilation.

## 8.13 The Future of Roofing Equipment

Drones are increasingly being used in roofing for surveying and inspections. They can provide high-quality images that can be used to thoroughly inspect roof conditions. Drone technology is especially useful for roofs that are difficult or unsafe to access. Drone cameras can also detect problems that might not be visible to the naked eye. In addition, infrared cameras attached to drones can capture thermographic imagery, which can detect roof anomalies. Drones are commonly used today, but their technology is constantly evolving. The Federal Aviation Administration regulates the use of drones, and a certified remote pilot must oversee their operation.

An advanced roll-forming system uses a computer to make and cut metal roof panels to their desired length. It may also be possible to apply measurements taken by drones directly to these machines for fabrication, making the whole process safer and faster.

Robotics are commonly used today for manufacturing roofing materials and equipment and may also be adapted for roof applications.

Tracking roofing materials, equipment and other assets by Radio Frequency Identification (RFID) is being adopted slowly by the roofing industry. While RFID is used today mostly on large projects, the technology has proven to be successful in tracking and managing materials and power tools.

Healthy Lifestyle (HLS) and Quality

Managing quality control on a roofing project typically includes the following:

- The roofing foreman and superintendent, who work for the roofing contractor and are responsible for the proper installation of the roof system

- The roofing material manufacturer's technical representative, when the manufacturer issues a warranty for the materials

- An independent quality assurance inspector, usually a roof consultant, who works directly for the owner

- A building code inspector who focuses on code compliance for the project

This structure has been used for quality control for the last decade or so. Previously – and still in some cases today – the primary source of quality control has been the roofing foreman and superintendent. However, due to changes in materials, methodologies and technology, QC measures vary broadly from foreman to foreman and job to job. And while the foreman is critical to the QA process, he or she has a crew and lots of daily duties to manage. As a result, the industry has adopted more consistent QC/QA practices, involving more than a single person in the effort.

Manufacturers' field representatives usually understand the installed product's capabilities and limitations. They are an excellent source for assisting with unusual roof conditions and help develop standard protocols for using the manufacturer's products.

The manufacturer's field technician also provides an extra set of eyes on the installation of the roof system.

However, it is important to note that the technical representative is only present during the installation of the manufacturer's specific product and is most interested in the product manufactured by the company he or she represents. They typically are not present throughout the entire project.

The building owner often hires an independent quality assurance inspector – most often a roof consultant – to ensure compliance with project specifications throughout the construction process. A consultant might also create the roof specifications, solicit bids and act as a construction manager. He or she could be involved in capacities ranging from conducting weekly inspections to inspections after phases of the project are completed to conducting multiple inspections per week.

A code enforcement official is responsible for enforcing building codes for the jurisdiction they represent. They are not directly involved in the construction process and do not control quality. A code official may provide input on how to comply with code requirements. However, due to the vast amount of roof systems available and the code officials' responsibility for enforcing codes very broadly, they may not be very knowledgeable about specific roofing products and details.

An independent quality control inspector becomes an integral part of the project team and is responsible not only for the identification of quality issues but also for resolving them. They have the authority to stop work if needed to make necessary changes. They look at the process and individual deficiencies to determine whether an issue is isolated or systemic.

While the older model of QA/QC mostly relied on a single person–the roofing foreman or superintendent–most people now understand that good QA/QC takes a team, usually composed of:

- The site manager (foreman)

- The construction manager or superintendent

- The project manager

- An inspector or manager

- A trainer

Quality control focuses on planning, teaching, training, coaching and enforcing. Quality assurance focuses on inspection, analysis, evaluation, documentation and communication.

To continuously maintain quality on a job, it is essential to ask the following questions:

- What are we doing?

- How are we doing it?

- Who is doing what part of it?

- How do we know if it is being done correctly?

- How do we get things back on track when something is done incorrectly?

- Who does what documentation?

- How do we do it better?

Establishing the standard of work is the first step in developing a QA/QC plan. Multiple organizations, as shown in Figure 9.3.1, either create standards (e.g., the International Code Council and FM Global), test against existing standards (e.g., UL) or recommend best practices (e.g., the National Roofing Contractors Association).

Establishing the right standards to follow depends on the geographical location of the project, the owner's requirements and cost constraints. It is important to establish the project's standards before beginning the project. Standards must be communicated to all workers, and the workers must be trained accordingly. Standards are reviewed and improved over time based on changes to market conditions, changes to industry conditions, development of new products and advancements in technology and safety.

Continuous improvement can be thought of as a cycle that includes:

- Determining the standard
- Planning for communication of and training for the standard

- Implementing the standard

- Monitoring and measuring against the standard

- Review the successes and failures of following the standard

- Determining if changes to the standard are necessary

Quality planning and coordinating meetings are critical but are often overlooked in managing a project. These meetings set the course for the entire project. At these meetings, the QA/QC team is identified, and their responsibilities are established. These meetings help identify the critical activities that need monitoring and the sampling interval for inspections and testing. These meetings help to develop a plan for communications, inspection findings and documentation, and also help to establish training needs for the crew.

It is essential to review the approved product submittals at the beginning of the project. This information will be needed to perform routine inspections at the start of and during construction to ensure that the right products are being installed according to the specifications. The manufacturer's product data sheets must also be reviewed for test results, and a test method and interval of testing should be established to ensure the products on the job meeting meet the standards specified in the product data sheets. The manufacturer's storage and handling requirements also need to be reviewed. Materials should be sampled upon delivery and at regular intervals, and readings documented and retained in the event a warranty claim is made or problems arise.

Preconstruction meetings must be held with the foreman, construction manager, project manager and QA/QC inspector. Drawings and specifications must be reviewed during this meeting and must be understood by all stakeholders. If possible, these meetings should be held on-site to help correlate the drawings and specifications with existing conditions and identify and rectify any issues before the start of the project. The foreman and construction manager should each have a copy of the most current drawings, specifications and any addendums. Tasks in progress must be routinely inspected and compared with the document requirements. Inspections should be conducted daily by the foreman and project manager and with the QA/QC inspector as often as predetermined during the preconstruction meeting. Any deficiencies must be addressed, and the daily QA/QC inspector must document remediation measures.

Safety and QA/QC go hand in hand. Clients view safety as a critical factor when selecting a high-performing contractor. A safe project is an organized project. It is essential to break down key activities into tasks and to identify hazards for each task.

The ideal solution is to address any potential hazards before the project begins. If the hazard cannot be eliminated, a plan should be created to reduce its impact. The safety plan must be re-evaluated regularly as conditions change.

Housekeeping is also related to QA/QC in several ways. It is a key visual indicator to the client about the quality of the work being done. Expectations for good housekeeping should be communicated to all workers on site. The lack of organization and cleanliness can quickly transition from an aesthetic issue to a safety concern. Debris from the roof could fall on passersby, for example, or cause slips and falls on the roof. Debris on the roof can also affect the productivity of the workers. Newly installed products might be damaged or otherwise affected by the lack of maintenance during construction.

Standards must be communicated to the QA/QC installation teams before the work begins. The team must know what the desired finished product will look like as well as the process of installation to be followed. The tasks to be performed and relevant standards can form a checklist that the foreman, construction manager and QA/QC inspector can review daily. Deficiencies, however small, must never be ignored. Records of deficiencies must be kept so that the crew's behavior can be evaluated. Training efforts can then be focused on areas that need attention. Doing something right the first time costs a lot less than reworking it.

## 9.4 Typical Repairs

Examples of typical QA/QC reports include the following:

1. Daily Construction Manager (CM) Report. This report includes a summary of daily activities that are communicated to the key stakeholders. The report includes these sections:

    a. Scheduled activities completed

    b. General description of work in progress

    c. Problems encountered, actions taken and delays, if any

    d. Meetings held, participants and decisions made

    e. Subcontractors, suppliers and company crew on site

    f. Visitors and purpose of the visits

    g. General remarks

    h. Ideas for improvement

    i. Weather conditions

    j. Photographs as appropriate

An example of the daily CM report completed by a roofing contractor is shown in Figure 9.4.1.

| Report date: | 4/9/21 (created 4/9/21) |
|---|---|
| Start time: | 4/9/21 07:00 am |
| Hours worked: | 9.00 |
| Number of men: | 4 |
| Squares installed: | 0.00 |
| Flashings: | No |
| Daily tie-in: | No |
| Fire watch: | No |
| Housekeeping performed: | No |
| Material look-ahead performed: | No |
| Inspection performed: | No |
| CERTA cards checked: | No |
| Details: | Crew came an probed roof an finished installing gutters an downspouts an counterflashings |

| Project dates: | 2/17/21 to 4/16/21 |
|---|---|
| Man days used: | 184.75 443.00 (258.25 remaining) |
| Daily total man hours: | 1478 |
| Squares completed: | 252 / 277.00 (25 remaining) |

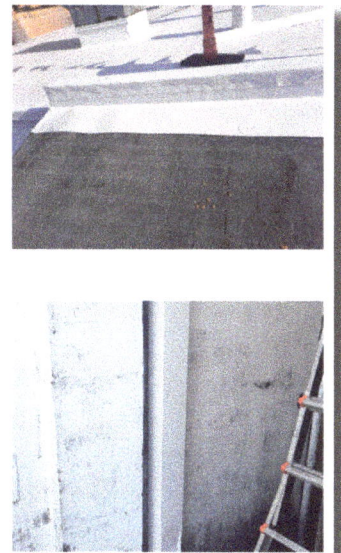

2. Preparatory Site Inspection Report. This can be part of the daily CM report or be a stand-alone document. It is distributed by the CM and includes the following:

a. Determination of whether any required prior work is completed correctly before commencing any new activity or task

b. Verification of field measurements

c. Assurance of the availability and quality of required materials

d. Identification of any nonconformance to the requirements necessary for a new ask to begin

e. Identification of any potential problems

3. Task Ready Inspection Report. This is also prepared by the CM and can b either a part of the daily report or a stand-alone report. For each task, the CM or the QA/QC inspector performs job-ready quality inspections to ensure that work activities being only when appropriate. The inspection also is used to verify that the job site conditions meet the project's quality requirements.

4. Work-in Progress Inspection Report. This report is either part of the daily report or a stand-alone report. For each task, the CM or QA/QC inspector performs an initial work-in-progress inspection when the first portion of each activity has been completed. Following this, an additional inspection is conducted to ensure that the ongoing and completed activities continue in conformance with the project's quality requirements.

5. Quality Control/Assurance Report. The QA/QC inspector conducts a routine project inspection to observe site conditions, tasks and activities and to follow up on daily QC inspections. This report typically includes:

      a. Commentary on who was present during the inspection

      b. Commentary on what tasks were completed and whether they were accepted

      c. Commentary on what tasks are in progress and whether they were accepted

      d. Acceptance or non-acceptance of materials to be installed

      e. Acceptance or non-acceptance of housekeeping

      f. Acceptance or non-acceptance of safety procedures used on the job

      g. Identification of remedial efforts – who, what and when

      h. Identification of corrected issues

      i. Identification of outstanding issues to be addressed – who, what and when

Typically, a monthly quality control report is produced for projects that run for several months. They act as a composite of weekly reports. An example of a QA/QC report is shown in Figure 9.4.2. Notice that the report includes discussions of safety, housekeeping and project management. Color-coded tabs for each category show the different activities. At the end of the report, a star rating shows how well the project is doing in terms of quality and also shows areas that need improvement.

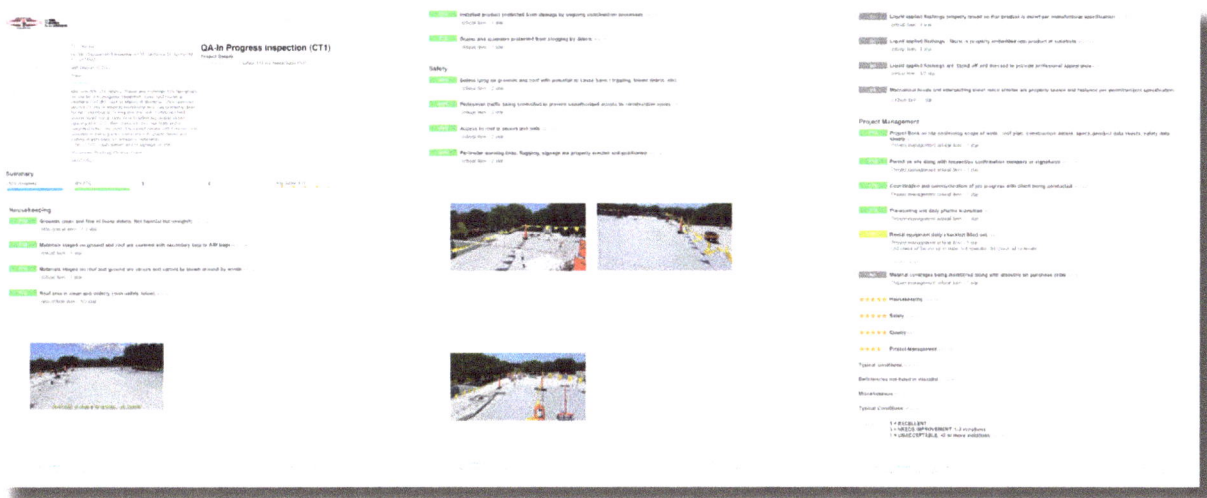

Deficiencies noted in the reports can provide insight into worker behavior patterns and allow for more focused training. It is always a good idea to use industry professionals to develop and implement training materials, and it is also a good idea to have recognition or rewards programs in place, for example, to recognize a crew with consistently high QC scores, overall good performance, the most improvement, the shortest punch list, and so on.

Lessons learned are also a valuable tool for crew education. Deficiencies should be discussed during weekly or monthly meetings, where focusing on quality ideals and training for them is a great way to identify and create new leaders who will bolster and maintain the company's quality control standards.

There are a number of software platforms available that aid in creating and implementing sound quality control inspections, documentation and response management. Some of the more common ones are:

- FTQ360, a web-based program that can be easily be tailored to different disciplines in the construction industry. It can be used for building checklists and checkpoints and is customizable to fit individual needs. It can also easily import and export data and be used with other construction management software.

- INTELEX Quality Management Software is a program specific to quality control and is used in the construction and manufacturing industries.

- HCSS Construction Management Software is also web-based and customizable to fit individual needs.

- Pro-Core Construction Management Software is a commonly used project management platform and is all-inclusive with modules for quality assurance and safety.

When selecting software, it is always important to be sure the software aligns with the company's objectives. Good software will provide planning, tracking, reporting and analytical functions that will help create a solid QA/QC plan.

## 9.7 Testing

Tests associated with low-slope roof system installation include:

- Fastener or adhesive pull tests to measure wind uplift performance

- Peel tests for sealants or coatings to measure adhesion and elongation performance

- Moisture content surveys of cementitious roof decks or of existing roof systems in a recovery roof application to be sure the moisture content is acceptable

- Compressive strength tests for lightweight concrete decks

- Test to measure mil thickness of membranes, sheet metal, adhesives or coatings to be sure they meet specifications

A fastener or adhesive pull test is typically performed before a project begins to confirm the structural integrity of the roof deck. These tests are sometimes performed during the application of the roof system if issues of deck integrity are suspected in localized areas of the roof. The responsible party then takes remedial action and documents the results. A simple gauge that engages the head of the fastener and pulls it until it fails. The device is attached to a 2'x2' area of roof insulation that is attached to the deck. The meter on the device shows the pressure applied. Fasteners and adhesives need to meet minimum pressure per square inch standards as specified for the project.

A peel test is used to test the adhesion strength between two materials. The roof coating adhesion test, shown in Figure9.7.1, involves using a fabric underneath the coating that is pulled using the hand-held scale. Another type of peel test, shown in Figure 9.7.2, is a 90-degree test that measures the adhesion and elongation of joint sealants. A part of the sealant is marked and one end is cut. The cut end is then pulled at a 90-degree angle to the wall.

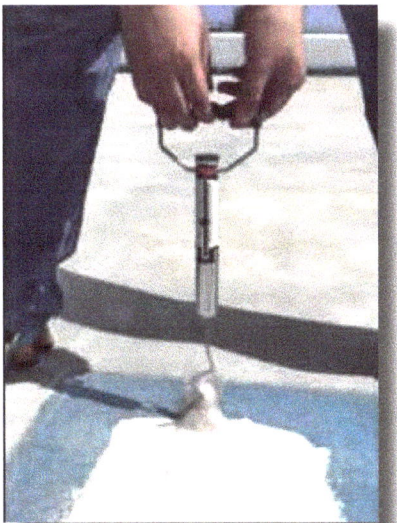

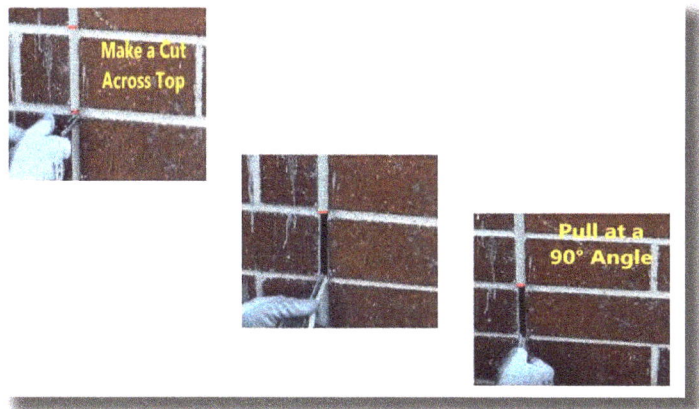

Traditionally, a plastic sheet is taped over a small area of the deck to test the moisture content of a concrete deck. This method was believed to allow moisture to rise from the deck to the underside of the plastic sheet. However, a more reliable test method was developed, shown in Figure 9.7.3, where an electrical isotope impedance meter is drilled into the deck to measure moisture content at specific depths.

Moisture testing within the roof system is done using thermographic imagery or through isotopic meters, as shown in Figure 9.7.4. It is essential to remember that thermographs might not always be accurate, as environmental factors and the reflectivity of surrounding objects can easily affect them. Workers using isotopic meters must have a license. These meters come with a transmitter and receiver and measure the amount of hydrogen molecules in the system, indicating the presence of water.

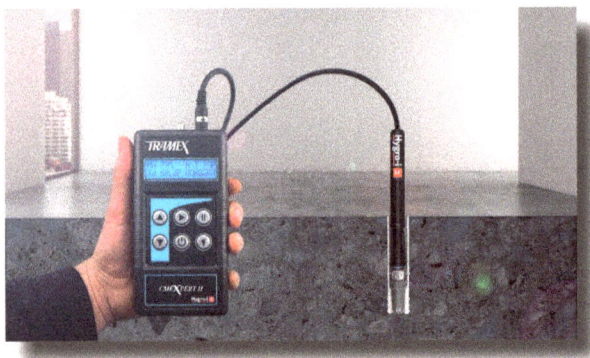

Figure 9.7.3 – Electrical Isotope Impedance Meter

Figure 9.7.4 – Isotropic Impedance Meter